The Awesome
Life of a
Physics Teacher

The Awesome Life of a Physics Teacher

Paul G. Hewitt

City College of San Francisco, USA

World Scientific

NEW JERSEY • LONDON • SINGAPORE • BEIJING • SHANGHAI • TAIPEI • CHENNAI

Published by

World Scientific Publishing Co. Pte. Ltd.

5 Toh Tuck Link, Singapore 596224

USA office: 27 Warren Street, Suite 401-402, Hackensack, NJ 07601

UK office: 57 Shelton Street, Covent Garden, London WC2H 9HE

British Library Cataloguing-in-Publication Data
A catalogue record for this book is available from the British Library.

THE AWESOME LIFE OF A PHYSICS TEACHER

ISBN 978-981-12-9865-3 (hardcover)
ISBN 978-981-12-9866-0 (ebook for institutions)
ISBN 978-981-12-9867-7 (ebook for individuals)

For any available supplementary material, please visit
https://www.worldscientific.com/worldscibooks/10.1142/13999#t=suppl

Desk Editor: Rhaimie Wahap

Typeset by Stallion Press and Lillian Lee Hewitt
Email: enquiries@stallionpress.com

Dedicated to friends who most influenced my life —
Kenny Isaacs, Ernie Brown, Burl Grey, Huey Johnson,
Ken Ford, and especially my wife Lillian.

INTRODUCTION

Among the giants in the physics world in the twentieth century who inspired me were Albert Einstein and Richard Feynman. I had the honor and pleasure of meeting Feynman in 1987 at a Southern California American Association of Physics Teacher meeting at La Cañada High School near Pasadena. I was part of a panel of educators who addressed the question: What kind of physics should be taught in American high schools? I arrived early and spotted Feynman standing alone in the hall. This was a common situation for him; teachers were uncomfortable in his presence. I was not, especially since I was told that he had just looked through the brand new high-school version of my *Conceptual Physics*. I was pleased but not surprised that he did more than look through it. He read it and had quite a bit to say about it. This confirmed his reputation for always "doing his homework" when contributing to such meetings. He suggested some improvements and I implemented them in the second edition of my book.

Being on the panel was a flabbergasting experience because I was debating my hero. Feynman's position was that physics should *not* be taught in high schools but postponed until college. Why? He felt that high-school teachers would screw it up, mainly because they lacked a passion for physics and knowledge of it as well. In short, students would benefit by a first college course with correct physics and the math taught right. My position was that physics should first be introduced with no math problem-solving, which could be postponed to a second course. I said that we need to change the "killer-course" reputation of physics by teaching it conceptually in a non-mathematical language so that it could join reading and writing as central basics in the educational mainstream—in short, physics for all. Feynman asked where qualified teachers would come from to teach a physics course suitable for all.

Sadly, this was Feynman's last public appearance. He died shortly after of abdominal cancer he had fought for years. The panel event was televised, but allegedly, the videotape was botched. To my knowledge, no video recording of the meeting exists.

In response to Feynman's question, I hope that my story will inspire others to become those qualified teachers.

If a learner's first course in physics is delightful,
the rigor of a second course will be welcomed!

CONTENTS

ONE

The Early Years

W hen I was six years old or so my father rented a rowboat from the pier at Revere Beach (the Coney Island of Massachusetts) and brought me out to sea so he could drown me. My sense of worth was pretty low, so I was okay with that. And he had my sympathy for being badgered by my mother to pay more attention to me. So my unfortunate father, much preferring to spend his Saturdays with his friends at various bars, succumbed to taking me fishing. Of course he had no intention of drowning me, and the time we spent together in the rowboat off the shore went well. I knew my father loved me in his way, and I certainly loved him. It didn't matter that he never played ball with me, took me fishing, except this one time, or buddied with me as most fathers do with their sons. One reason was that Ted Frieswyk, who lived up the street from us in Saugus, was eager to do those things

with me. The Frieswyks had one daughter, and I became the "missing son" for Ted. That relationship, which wasn't a problem for my father, had no downsides whatever, and I had a varied and interesting childhood.

Faith Isabel Nee

Paul Gordon Hewitt

Back in 1929, more than a year before I was born, Faith Nee laughed at my father when he said he was going to marry her. That was during a dance at Beachview Ballroom in Revere Beach. My mother was a lovely seventeen-year-old girl that my dad, then twenty-one, wanted for his bride. In my dad's favor wasn't so much his good looks—he looked like Fred Astaire, high forehead and all, but like my mother was a terrific dancer. My mother told me much later that she wasn't impressed with his confident wishes and particularly the way he shooed other suitors away as he sat on the front porch of her home in Everett, Massachusetts. Her mother, later to be my grandmother, learned that her daughter's persistent suitor had a high-school education and was a good prospective son-in-law. She encouraged her daughter to accept his brash advances. Having three daughters of her own and two of her sister's daughters living with her, parceling off young Faith to a good man was prudent.

On November 30, 1930, my parents were married. My dad had already secured a job at the gas and electric company in Revere, where he worked for the rest of his life. They settled in nearby Winthrop, my dad's hometown, and on December 3, 1931, slightly more than a year after a small wedding ceremony had their first child—me. I have the same name as my dad but have seldom used the Junior (Jr.) that goes with it. My coming into the world was the end of Faith Nee's ballroom days—the end of the prettiest and most sought-after girls on the ballroom floor. Instead she had the task of caring for a child with a man she gradually came to love.

If I were born a couple of years after their marriage, allowing them time to continue their wonderful times at the Beachview and nearby Oceanview Ballrooms, I think that I might have been more welcome. My mother always seemed unsatisfied with me, with a mild resentment that lasted all her life. This wasn't clear to me until years later, and besides, how else could it be? I never suffered hardships that other kids endured, and I knew my parents loved me, so life was good. Good, but not good good. Harmony between my parents was strained throughout the years, underscored by both of them being sexually attracted to others which led to episodes of promiscuity that resulted in heated quarrels.

All my childhood years were in Massachusetts. My earliest memories are from the age of three, when we lived in a rented home on Grover Street in Everett. At the end of the street was a dairy farm and cows were often herded along the street. I remember watching from the balcony and was curious to see some of the cows playing piggyback as they walked along the unpaved street. Ah, animals liked to play. I was also curious about money. When I was a bit older, my mother would send me to the store with a single coin and a note listing what was needed. I was puzzled that I'd give one coin to the man and he would give me a bag of things plus more coins as change. A curious system, I thought.

I began school at the age of four, three months shy of five, and was promoted on trial to the first grade at Lafayette Elementary School in Everett. Much of this schooling was confusing to me. I remember my teacher, Miss Tragone. I mispronounced her name as Mister Gone, wondering why women teachers were addressed as Mister. Early in the year she asked the class to draw a square, then in the window of the square draw a letter A. To me, *square* meant downtown Everett Square, where my mother often took me. We used to look in the large window in Kresge's department store in the middle of the square. So I made a sketch of Everett Square, showing the front of Kresge's building and all, and in its display window I drew a large letter A.

20 Fairview Avenue in Saugus

Perhaps Mister Gone thought I was being mischievous, for she kept me after class.

After we moved from Everett to neighboring Saugus, my second-grade teacher at Cliftondale School was Miss Graves. In an art lesson the class was advised to make a drawing of President George Washington giving a speech to a line of troops. President Washington and the troops were to be drawn all standing. My drawing got Miss Graves' special attention and referred me to Miss Kellogg, the art teacher for all schools in Saugus. Rather than drawing rows of standing soldiers the same size, I instead drew them successively smaller the farther away from President Washington. Unusual for a second grader, the perspective was "correct." I was happy that rather than criticism, the teachers praised my grasp of perspective. Nothing came of it, but I much later found I scored highest in spatial reasoning on intelligence exams.

During third grade my friend Eddy Lyons and I arrived late after lunch too many times. We were both sent to the principal, Mrs. Beckman, who taught fourth grade. I was the skinniest kid in the school and Mrs. Beckman saw this as an opportunity to prompt me to eat more. In front of the fourth-grade class she compared Eddy's rosy and full cheeks to my sallow sunken ones. She also held up my arm to show how skinny it was compared with Eddy's arm—as if I needed to be reminded that I was skinny and should eat more! That humiliation diminished me for the early years of my life. No amount of eating could affect my skinniness. I continued my education at Sweetser Junior High School, where I encountered my first male teacher, whose name I forget. In those days most teachers were single women.

The first yum Christmas present I remember was my first bicycle. It was a used one that dad painted blue. It had wooden wheels. That's right: wooden.

There was a time when rubber tires with their tubes were mounted on rims made of wood—which soon were replaced with metal rims. Our home in Saugus was almost at the top of a hill, which to children seemed very steep. How I envied friends who lived in the flat areas of town, for they could ride their bikes from their garage and simply peddle effortlessly back, all on level ground. No pushing your bike up the hill whenever you went riding. So I didn't get much use of my first bicycle.

Margie, Dave, and me in back yard snow

My childhood in the 1930s into the 1940s included sister Marjorie and brother David. Margie, as we called her, was born in 1933 and Dave in 1937. Girls hung out with girls, and boys with boys. So my sister and I had different friends, although we often played paper dolls, "papes" as we called them. That's when I wasn't drawing comic strips or making model airplanes. I relished the skills that were a part of building model airplanes. Thin strips of balsa wood were glued to form wings and a fuselage, which were then covered with thin paper. The final touch was painting the finished model with shellac. My pride was a two-engine double-fuselage P-38, a World War-II fighter airplane that I painted a silver-grey. I hung this and other beauties on the ceiling of the bedroom that brother Dave and I shared. Years later, Dave told me that on occasions when he was annoyed with me he'd poke holes into the P-38. I had never noticed this. For the most part, my five-year younger brother and I got along just fine.

Me and my siblings

Margie, Dave, and I all enjoyed our childhoods, for there were ample woods and a saltwater marsh not too far away where we would swim in summer and ice skate in winter. My sign-painting skills morphed into silkscreen printing. Dave skipped college and successfully went into engineering, specializing in electronic circuit boards before the advent of chips.

Margie attended a Christian college in upper New York—Houghton College. Years later, she earned a PhD in religion at Claremont Graduate University in California. She became a world-class theologian specializing in process theology, which grows from the fundamental work of Alfred North Whitehead, a noted early 20th century mathematician, philosopher, and scientist. Authoring several books in her field, Margie is the religious one in the family.

My youngest brother, Stephen, was born when I was seventeen in 1948. Just as I spent a lot of boyhood time drawing comic strips, Steve in his young years drew maps of an idyllic and tropical part of the world he fantasized living in some day. Years later he traveled south with his friends to tropical countries on motorcycles, found his paradise, and today lives his childhood dreams in beautiful Costa Rica. While there, and after marriage, he turned 200 acres of hillside into a coffee plantation while maintaining a maritime career. He built a beautiful spacious home that I visited annually in later years, which also hosted family reunions. He retired as a Captain.

Stephen and dad in our back yard

I was barely ten years old when the Japanese bombed Pearl Harbor, the beginning of World War II for the United States. I remember our family gathered around the radio to hear President Franklin Delano Roosevelt give his day-of-infamy speech. Food was soon rationed, with different designations for meat and groceries. Milk continued to be delivered daily and swill and trash were picked up weekly. Swill was garbage suitable for pig feed and was dumped into a cylindrical tub with its top at ground level. Trash was put into another container.

Everyone read about the war in the newspapers and listened to the radio. Being curious, I asked my mother why the Germans and the Americans were so different. Wouldn't some ideas of both sides be acceptable enough to not be fighting? Like enough so that soldiers on both sides could be playing checkers or chess with one another instead of fighting to their deaths? She replied that it didn't work that way because each side was convinced they were right. Sadly, it was an "us" versus "them" world. This bothered me for many years—actually, to this day.

Our family belonged to the Cliftondale Congregational Church. We kids attended Sunday school, where the values of kindness and compassion were taught rather than any religious indoctrination. A bonus of Sunday school was being a bit awed by how pretty all the girls were, wearing dresses and ribbons instead of play clothes. When our parents joined the church, they decided to have us baptized. So one Sunday we all went up to the front, and our wonderful pastor, Mr. McDuffee, baptized us. He had the kindest face a person could have; I loved him and his sermons. I was so disappointed when he retired! It was hard getting used to Mr. Cross, the younger man who replaced him. One time when I was alone with Mr. Cross I asked if there really was a God. His reply was simple: Albert Einstein believed in God. That did it for me. Knowing that Einstein was the smartest person in the world, I didn't question God's existence for many years.

Mom brought us up with an important lesson: Most people are decent. Not all, as daily news too often reminds us, but *most*. Her belief is one that has followed me all through life. It is something to remember, especially when attention is focused on the misdeeds of others. Most people are decent.

I also joined Troop 62 of the Boy Scouts at our family's Congregational Church. I had the customary brown uniform and found the experience a worthwhile social activity. A big highlight of being a scout was a summer camp week at Camp Nihan, somewhere north of Saugus. Being away from family in safe and interesting surroundings was most enjoyable. The activities of scouting were a lot of fun—building fires without matches, identifying local trees by their shapes and leaves, making maps, learning semaphore (signaling by flag positions) and the Morse code. My highest rank was First Class, with a few merit badges. I was elected patrol leader, due to disagreements between two popular candidates that resulted in more votes to me.

In 1945 I began my first year at Saugus High School, where I continued my habit of sketching things that I had begun in elementary school. The school guidance counselor, aware of my art ability told me that I wouldn't have to take academic courses in science and mathematics, but rather I could take easier courses and continue in art. That's right—I wouldn't have to *endure* the harder courses. So the only math and science classes I took were Basic Math for Boys and General Science in the ninth grade. I fortunately took typing (that became enormously useful), along with metal and wood shops, and other courses. My favorite course was Commercial Geography, due to its inspiring teacher, Mr. Walter Blossom. I prided myself for never having to take homework home, which I completed within the study sessions each day.

At Saugus High I made drawings for school publications. Comic strips were my interest, and I drew one of a hockey game where the goalie gets all his teeth knocked out—very funny, I thought. Mr. James Burns, the faculty advisor, rejected it. He suggested that I draw something more akin to school, for example, showing ways in which books get lost. So very reluctantly I drew a strip showing just that. Thinking it was boring, I made a disclaimer and lettered, "don't blame me, I only drew it." So what resulted was, "How Books

Get Lost, by Paul Hewitt—but don't blame me, I only drew it." Friends saw it as a short poem and thus my "Hewitt Drew it!" moniker was born.

For daily school lunches Ma made sandwiches for me, all wrapped in waxed paper. When she asked what my favorite sandwich was, my reply was tuna fish. Oh oh . . . *every day* it was tuna fish, tuna fish, and more tuna fish. Ma overdid it. It took some years before I could enjoy a tuna fish sandwich again. The message is, don't overdo a good thing.

Delivering Sunday newspapers with Dave

A very good thing was getting a job delivering newspapers on a Sunday paper route that included parts of Cliftondale and North Revere. My brother Dave helped in this money-earning effort. With earnings from the route I first purchased a red, black, and white brand-new beautiful Sears Roebuck beauty with two headlamps, which to me was the crown jewel of bikes. But as I said earlier, having to push it up the hill lessened its use. The same for orange crates mounted on slabs of wood with roller-skate wheels fastened underneath, long before the advent of skateboards. My second major purchase was a washing machine for my mom. My third was buying a 1934 Plymouth automobile for seventy-five dollars.

Pets were a part of growing up. Trixie was my pet cat. I loved Trixie, and when he'd venture away for more than a couple of days I missed him badly.

My cat Trixie

I sang love songs of the day to my Trixie, who lived until I was about sixteen years old. My only dog was Flicka, an Irish Setter that my parents purchased for me at about age nine. My love for Flicka was as much as my love for Trixie. My parents didn't know that puppies should get canine distemper shots, and sadly, Flicka died of distemper at the age of nine months. Both their graves were dug beyond the hill in our back yard. I made daily visits for some years. My sis and her friend Barbara followed me one late afternoon and overheard my singing ritual at the gravesite. I felt a bit intruded upon when they confessed to their transgression. Soon after, homes were built on the gravesites of Trixie and Flicka.

In the surrounding woods there were blueberries, which my mom and other moms converted into yummy blueberry pies. Just beyond the top of our dead-end street and downhill was an enchanting swamp where purple irises and lady slippers grew. The swamp had frogs and we delighted in watching pollywogs turn into these creatures. There were also vines that we could swing from in Tarzan-like antics, especially from perches we'd build among tree branches. Building tree houses was a great pleasure. We also played war games in the woods and made war toys. My neighbor Georgie and I were the best crafters. We both made machine guns, his a simulated 50-caliber water-cooled one and mine a 30-caliber smaller one. His was painted silver and mine was painted in green and brown camouflage colors. Somehow mine got lost. Several days later in Georgie's basement I opened the furnace door to discard a candy wrapper and discovered green and brown painted bits of wood, the remains of my machine gun. I don't remember confronting Georgie about this. This was my earliest encounter with envy.

Most homes back then were heated with coal. A coal truck would dump a winter's supply in a basement area designed for coal storage. Dad showed me how to position coal in the furnace so that it would burn all night. The bottom of the furnace interior had a series of gratings that held the ashes and coal. Each section of grating had a gear in the front so that when one grating was rotated, others did the same, allowing ashes to fall below. A handle was used to rotate the gratings. After ashes fell to the bottom they were removed with a shovel. Dad taught me these things. But I was most fascinated with the gears. So I constructed a model on a piece of wood, with a pair of gears cut from paper with pins holding their centers. But instead of a pair of identical gears, I made them different sizes to test my hypothesis that rotating one could multiply the rotations of the other. I made one twice the size with twice the number of teeth as the other. Sure enough, when I rotated the large one through one rotation, the smaller one rotated twice. This excited me and I showed it to my dad. He simply said that idea had already been invented and told me that such gears were used in cars when changing speed. Shucks! For a while I thought I invented a new idea. I didn't pursue my interest in gears.

The criminal activities of my youth included raiding cherry and apple trees at night. Another was sneaking into neighbors' gardens and snitching tomatoes in broad daylight. And any wood needed for a tree house was not purchased at a lumberyard but was snitched wherever available. Instead of stealing, we more politely "snitched." My worst crime was dribbling gasoline down the middle of neighboring Cottage Street and then lighting a match to it. As luck would have it, I poured gasoline in the middle of the road rather on the side where cars were parked. I was aghast at the magnitude of the fire that resulted. Fire engines came and I ran for the nearby woods and hid for a couple of hours before returning home. Fortunately, the event was without consequence to any property. Speaking of fire, I look back at how I placed nice comfortable hay in the play area under the front porch of our home. It was certainly a tinderbox, and I was lucky that no fire ever resulted.

Another crime was committed when I was about sixteen. Friends Bud and Dick Smith lived near a small farm owned by Mr. Olge, a European immigrant with a thick accent. His field of corn was ripe, and we filled our arms with fresh corn. When Mr. Olge appeared, we ran up the neighboring dirt road to the Smith residence, carrying our booty. I looked back after reaching a safe distance and saw Mr. Olge looking at us without expression.

He was resigned, with his hands hanging at his sides. At that moment I realized the extent of this snitching—Mr. Olge was hurt. He had labored over that corn that was not ours. The corn in our arms was his. We had violated him. I felt very guilty. My snitching activities ended with that event.

Between my junior and senior year at Saugus High School I was employed at a shoe factory in neighboring Lynn. While working there I met a lifelong friend, Peter Theologou. Peter's passion was dancing, and he weaned me away from roller skating to ballroom dancing at the Oceanview Ballroom in Revere Beach—the same ballroom my parents had frequented years before. Peter coached me, and knowing I was shy with girls advised me to get on the dance floor immediately. If we dallied, I'd lose my nerve. So

Peter Theologou, fifty years later

up the stairs to the ballroom floor we went when a fast-paced Viennese waltz was playing. I asked the first girl who looked like she wanted to dance and stumbled through the quick steps required of this dance. As soon as the music finished, pausing for the next dance of the waltz set, she walked off the floor. I went where Peter was waiting for me, and he pushed me to another partner. The pace was slower, and I entered the world of ballroom dancing. As much as I loved roller-skating, I found a new activity to love. I'm grateful to Peter for expanding my social horizons. I've been a dancer ever since.

At age seventeen or eighteen I developed a stomachache that I thought was due to overeating peanuts. I was sick in bed and didn't want to see a doctor as my parents had advised because it would cost me fifteen dollars, more than I had saved. Such expenses were borne by me, not my parents. They summoned our family doctor, Doctor Weiner, who after a brief diagnosis declared that I had a burst appendix. An ambulance brought me to Saugus General Hospital where I underwent emergency surgery. That night a nurse sat by my bedside holding my hand. I was impressed, but I later found she was there to record the time of my death. But a new drug was given to me—tetracycline—which saved my life. A draining tube was inserted into my abdomen. A family friend whom I called "Aunt" Winnie was fretful that I wasn't eating enough food at home, so she insisted that part of my recovery be at her home in Everett. So off to Everett I went, which turned out to be a terrible idea. Her husband Duke returned from work one evening and was horsing around to entertain his kids and me. He jumped up and down on the bed singing "If I Knew You Were Comin' I'd've Baked a Cake," a pop song of the time. My laughter was uncontrollable and I begged him to stop lest I break my stitches. But my stitches burst, and to this day I have an ugly scar on my abdomen. Throughout the following years I've joked that the scar prevented me from becoming a Chippendale dancer.

Dennis Goslin was a pal whose dad was a car dealer. Dennis owned a spiffy 1942 maroon Pontiac convertible. He was envied for his beautiful car, which was one of the last to be built prior to World War II. One evening I was invited to join him and others in a drinking spree in his car. Dennis drove, and I don't recall that he drank. But three of us in the back seat did, whiskey followed by gulps of beer. This was most enjoyable. At last I understood why my dad was so happy when he drank. Standing in the back seat I yelled greetings to people we passed. But then things turned. I became deathly sick. Dennis pulled into a gas station and I rushed to the toilet and puked. My head ached like never before. Lying on the concrete floor I wished to die. I'd never imagined feeling so bad. The experience was horrible, exactly opposite to the initially happy reaction. It was my first experience at intoxication. I have never since experienced intoxication to that degree. Walking by a beer joint with its beer smell was something I avoided for many months. The result of that experience, plus dad's sad times with drinking, found me never having a drinking problem in all these years.

One day Dennis drove his Pontiac to visit me at my home. The car was now a bright chartreuse, for Dennis had it completely painted after an accident. The car was an absolute dreamboat. While I was marveling at its beauty, Dennis asked what I thought it was worth if he were to sell it. I told him he'd have no trouble getting $600 for it (which was a significant overestimate). He asked if I felt sure of that. When I replied yes, he

A dream come true: my 1942 Pontiac convertible

announced that he would sell it to me for $600. All I could think of was the great pleasure that would come from owning this amazing vehicle. I told him I'd gladly buy it. This meant I had to get a loan, for I didn't have that amount of money.

At about the same time my Uncle Laurie Baruffaldi, husband of my dad's sister Aunt Dot, took an interest in my art ability and announced to my parents and me that he'd like to sponsor me through art school. That's right, he'd pay the tuition and expenses of an art school education—normally a dream come true. But that meant I'd have to choose between becoming a professional artist later, or owning the Pontiac convertible now. The choice was an easy one. The convertible might get me a girlfriend. And even if I didn't I couldn't be happier if the car was mine. Of course, I chose the car.

The Pontiac was a beautiful vehicle, but I had to make it even more beautiful. Two searchlights were already mounted on the driver and passenger sides of the windshield. I installed a metal bird on the front of the hood with amber wings that lit up when the headlights were on. I also added a pair of chrome horns on one of the fenders and additional lights atop metal stems on the sides of the front fenders. Then colored lights around the back license plate. Interestingly, neither the horns nor the bird on the hood were ever stolen. Things were different back then. When somebody said my car looked like a Christmas tree, I stopped making additions.

I abhorred being the skinniest kid in school, from elementary through high school. I was also the victim of bullies who chased me time and time again. When they hit me, I didn't want to hit back for fear they'd just hit me harder. I remember one time running home yelling to my mother, who shocked me when she announced, in earshot of the bullies, that it was time to fight my own battles. But when I was almost thirteen, I saw a movie about boxing that was to change my life: *The Great John L.* It was about John L. Sullivan, the heavyweight boxing champion of the world. He had no worry of bullies! How I wished I could be like him. That Christmas my parents bought me two pairs of boxing gloves—a wonderful gift I would not have appreciated before being influenced by John L.

Lucky are those who find a passion in their teen years. I found two—boxing and roller skating, which I'll talk about in the next chapter.

What I did on bad-weather days

TWO

My Teens in the Ring and the Rink

A turning point in my life was seeing the movie, *The Great John L.*, the last bare-knuckle heavyweight boxing champion of the world who reigned just before the time of boxing gloves. One thing was certain: no bullies bothered him. John L. Sullivan was my hero, and more importantly, my role model. How I'd love to be like him! And for Christmas how lucky I was that mom and dad gave me two pairs of Everlast boxing gloves. Aha! I had something as important to do as creating comic strips. I'd learn to become a boxer!

My ducks were in a row, for as it happened, I lived two houses away from Eddie McCarthy, an amateur and later professional boxer. Eddie was a 135-pound expert in the ring and five years

Boxing lessons in Eddie's back yard

older than me. He was agreeable to sparring sessions, mainly in his back yard. At the time I weighed less than 100 pounds, and I tried desperately to top three digits. When no kids were around, I'd step on the drugstore weighing scale in Cliftondale Square. Being the skinniest kid in school, the most anguishing day of the year was when the school nurse weighed each student in turn, calling out the weights to the teacher who recorded them. All my classmates were privy to my pitifully low weight. How I wished to be normal!

At age 17

Boxing enabled me to progress from being a victim of bullies to self-assurance in fending for myself. A turning point occurred in a confrontation with classmate Jack Lapeley, a bully who used to chase me from school. An older boy came between us and set us up for a sort of duel. He placed a chip of wood on Jack's shoulder and with great determination I knocked it off. Jack at that point was supposed to hit me. He didn't.

Then the chip was placed on my shoulder. My fists were ready for him. Jack showed fright and refused to knock it off. He was aghast to see a courageous conviction he never expected. It was a great victory for me. Neither Jack nor any other bully ever hassled me again. The confidence I had in myself regarding physical confrontations remained all my life. I never hit a person with my bare fist, for I never had occasion to do so. Soon thereafter, however, I had a reputation for boxing. The neighborhood store owner called me "Killer." Friends at the local pool parlor called me "Champ." I relished my newfound reputation.

Eddie's celebrity status in boxing was exceeded by that of his good friend, Kenny Isaacs, a boxing icon from neighboring Lynn. I remember attending boxing bouts at Lynn Arena and cheering for Kenny. I wished I could be the kid in Kenny's corner who brought up the water bucket between rounds, for then I could get to know this local hero. As luck would have it, I got my wish.

Eddie went off to fight in the Korean War and before leaving he asked Kenny to take me under his wing. Two years later in 1948 Kenny was in *my* corner when I fought for the New England Amateur Athletic Union (NEAAU) boxing championship at Boston Arena. This was more than a dream come true—for I was in the ranks of my heroes.

Kenny Isaacs

Sparring with the likes of Eddie McCarthy and Kenny Isaacs for three years resulted in con-siderable boxing skills. Boxing with people my own weight, 112 pounds, was a snap. I easily won the NEAAU quarterfinals, and barely won the semifinals. My opponent, Mike Filipon, was the state flyweight champion and was known as a hard-hitting body puncher. In the third round he hit my midsection with an uppercut that was brutal. I winced with the pain, but turned that wince into a smile. If he had hit me again in the same place, I'd not have won that bout. Although he beat me up badly, I won by a unanimous decision.

I rested for about two to three hours in the dressing room before facing the finals. I was exhausted, for my skill in the ring was not matched with physical strength. While resting, I prayed to God saying that if I won, I would publicly honor him by kneeling in prayer in the center of the ring. Later I felt guilty about this. For one thing, what if my opponent was praying for the same? God would have a dilemma, choosing between us. I was a fool. A lesson learned: You don't make bargains with God!

The flyweight championship bout took place in the same evening, and I faced Ted St. Jean from Central Falls, Rhode Island. After the first round Kenny congratulated me for I looked good in the first round. Then I told Kenny that I had a stitch in my groin—the kind of pain that occurs when running or tiring too much. And the stitch wouldn't go away. I was done. In the second round I was pooped, and Ted punched me at will. My nose bled profusely, and I was a bloody mess when the referee stopped the fight. In

those days before television, such bouts were broadcast over the radio. My father was in the arena, but my mother was at home to hear on the radio that the referee "stopped the carnage." I returned home that night quite unhurt, but tired. I had won the silver medal, which I treasure today.

Aside from my cherished silver medal, the only evidence remaining of that never-forgotten event is the Boston *Globe* sports page photo of me being hit by my opponent with the caption "Lands Hard Right."

LANDS HARD RIGHT

Ted St. Jean (left) of Central Falls, R. I., is pictured landing a punch flush on the jaw of Paul Hewitt of Saugus in the first round of the finals of the 112-pound New England A.A.U. boxing championships at the Arena last night. St. Jean won in the second round by a technical knockout. (Photo by Joe Lynch, Post staff photographer.)

My high school principal called me to his office, and I expected a scolding for skipping class for the fight. Instead, he proudly held up the page with the photo and congratulated me. He added that I was a credit to the school and community. That was fine indeed.

If you asked me at that stage of life what my long-range goals were, I'd have answered that I wanted to win the national flyweight championship, then immediately retire from the ring (to not risk long term damages) and draw comic strips for a living. I'd been drawing comic strips since the age of eleven, and I still have many that survive today. Most are complete, inked and colored with colored pencils. I've always been inclined to complete tasks.

My boxing plans were interrupted in April in a warm-up bout for the national championship, again at Boston Arena. When a boxer signs up for a bout, he seldom knows who his opponent will be. In this case it was Franny Walsh, a previous opponent from Lynn in a "smoker" (a private exhibition with no winner or loser). Smokers were popular back then, usually sponsored by clubs of one kind or another. Boxing at smokers was "unofficial." Although this bout was official, I didn't expect a bad time with Franny. When the bell rang, he advanced quickly and knocked me cold with a single punch. My manager, Kenny Isaacs, said that my feet flew up and my head struck the canvas, beneath which was a plywood floor without padding. I remember nothing of this. I was unconscious for several minutes and woke up in the dressing room to see my dad looking down at me. He was quite scared. I vomited out the car window on the drive home. I suppose I experienced a mild brain concussion. Some days later I received a letter from the NEAAU, in response to a recent death in the ring, informing me that I was forbidden to box for one year. Although my interest in boxing never waned, that was my last boxing bout.

Something important that I gained from my boxing days occurred when I was in the dressing room at Salem Arena after a much-publicized boxing event. Promising fighter Tony Karas had just been badly beaten in a bout. Tony was born to be a fighter, urged on by his father, who had made a name for himself in earlier years. There was Tony, sitting on the table, bloody and bruised. What I most remember was his remark to those in the room. "There's always a better man—and I met that better man tonight." Wow! I was enormously impressed by the wisdom I saw in Tony. He reinforced my view that you don't have to be the best, but just be the best you can be (not to be confused with the army recruitment slogan). One who feels he or she must be the best is in a perilous position—for it can never be sustained. Better to take personal pride in your accomplishments, and if recognition comes to you, then even better—but not necessary.

My passion for boxing was matched by my passion for roller-skating. In a diary I say "I regret the day I become too old to roller skate." I was very much enthralled with roller skating, which began when late one afternoon a couple of friends and I rode our bicycles to nearby Revere Beach, the Coney Island of Massachusetts, and watched outside through the windows of the roller-skating rink, Revere Arena. We saw

Paul Ryan

a brand-new world. The next week we tried roller skating, and I was hooked. I met my lifelong best friend Paul Ryan at the rink. I also met girls, something I thought I was incapable of while I was in high school. The best part of all was that it didn't make any difference if you were skinny or not. If you skated well, girls said "yes" when you asked them to skate the dance numbers. To combat my skinniness, I nevertheless wore both a sweatshirt and a T-shirt beneath my shirt. That's right—underneath my shirt so that I'd bulk up my body. It was somewhat peeving when a partner would grab onto the sweatshirt instead of placing her hand on my shoulder, therefore knowing that I was trying to hide being skinny. But they never said anything about it. No girl asked why I was wearing so heavy a garment on a hot summer night. The shyness that accompanied being skinny resulted in never seeing any of the girls outside the rink. I didn't have enough confidence for that.

The wheels of roller skates were made of maple wood, as were the floors of skating rinks. (It would be decades before the advent of polyurethane wheels.) For racing, smaller diameter wheels provided stability and mobility.

I joined a Saturday morning racing team and enjoyed it. After a fall at the front of a line of racers I cracked my ribs a bit, and went back to free-style skating—executing rhythmic rubbery dancing leg moves across the floor as if falling down was impossible. Paul Ryan and I went through the cycle of first renting skates, then buying a pair, and saving our money to ultimately purchase the Cadillac of skates—Douglas Snyder precisions. How wonderful to wear the best roller skates in the world. Our skating world at Revere Arena was our teen paradise. Paul and I fantasized for years about those golden times.

New Year's Eve at Revere Arena was a godsend. That's when you could kiss all the girls with a cheery "Happy New Year!"—even the prettiest ones. The skating arena was absolute heaven at this time, but the once-per-year kissing evening was short lived for me. One of the things I had in common with my mother was cold sores. It seemed that I got these horrible things from my mom, who had terrible bouts with them. The sores were only on the lips, but with mom they extended to her nose. Obnoxious things! Later I learned that chemical changes in the body associated with the excitement of kissing activate the cold sore viruses. Only if a New Year's Eve could sneak up on me without notice would I be able to enjoy an evening of kissing all the girls. For succeeding New Year's Eves were accompanied by cold sores that spoiled everything. But it was super great that one time.

Kissing pretty girls was wonderful. But kissing could lead to what all teens wondered about—sex, which I didn't experience in my roller-skating days. From early childhood I had only misconceptions about sex. For example, my friends and I thought that the number of children in a household was equal to the number of times their parents "did it." We thought that "queers" were guys too ugly for the opposite sex. We knew about the birds and bees, but any extension to humans was a no-no, certainly not to be discussed in school classrooms. Planned families were uncommon in those years. The number of children in a family (later including my own) was often the result of contraception gone wrong. A teen neighbor became pregnant and we later learned that she was scooted off to a home for unwed mothers in Boston to avoid family disgrace. Having a child out of marriage was a dishonor not only to the girl and her family, but to the neighborhood. We hoped that the home found caring parents to adopt the baby.

More serious was the life of shame endured by one of my best friends, Ernie Brown, whom I first met in 1950. Not until he was inducted into the army at the end of World War II did he discover that he was illegitimate. Army records indicated that his Aunt Elsie was in fact his biological mother. "Brown" was a made-up name given to him by his grandfather. A baby born out of wedlock was a mistake, a nonperson, then called the offensive term "bastard." Ernie endured that stigma all his adult life until he died at the age of eighty. In fear

Ernie Brown

of bringing another bastard into the world, he lived a life of celibacy. That's the way it was back then.

Fast-forward to the twenty-first century, where ironically, single motherhood in America is often *celebrated*! Single moms are now common, with little of no shame associated with having babies out of wedlock. Low-income moms are subsidized, a policy that sometimes finds children having and raising children—all in a dad-absent home. Are kids without dads becoming the new norm? As an upside, at least they don't suffer the stigma that my friend Ernie endured.

THREE

Earning a Buck

My adventures at Revere Beach were not limited to the awesome world of roller skating. For several summers my dad worked at the Hippodrome, the famed merry-go-round of flying horses that children loved to ride. My brother and sister relished the free rides when dad was the ticket collector. Dad had bigger plans. He and two partners rented a square lot between an ice cream concession and a sandwich shop located away from the high rent center of the beach. They went into hock to purchase a gasoline powered mini train and installed train tracks to fit the open lot.

Class of 1949

Very unfortunately, what they did not know was that three other kiddie train rides appeared that same summer! My dad and his friends never recouped their investment. This was a tragic coincidence that devastated my dad and his partners—a horrible and undeserved blow.

I sensed an economic opportunity. What Revere Beach did *not* have was a carnival type high striker—the tall vertical wooden shaft that invites participants to strike a sledgehammer upon the end of a lever and send a missile upward toward a bell at the top. For each ring of the bell, the player gets a cigar. These were popular at various carnivals, and I was convinced that a high striker would be a hit at the beach. So alongside the ticket booth for the train ride, I placed my self-built high striker, with 30 percent of each evening's proceeds going to the landowner that operated the adjacent ice cream stand.

The two most festive days at Revere Beach were the Fourth of July and Labor Day in September. These two dates marked the beginning and end of the summer season. At age seventeen I rushed to build and paint the high striker in time for the Fourth of July weekend of 1949. It was a wooden board some thirty feet high with sections painted from the bottom to the top: Cream Puff, Pip Squeak, Balloon Ass, Lover, Almost There, and Superman. Initially I used a metal cable to carry the missile, but that had the effect of turning the contraption into a drawn bow. So with my Uncle Fleet's help we reinforced the back of the column and fashioned a metal track to guide the missile up toward the bell. I designed the missile and the fulcrum and had both manufactured at a machine shop, marveling at the machinery used. Upon the fulcrum was the lever, a six-inch by six-inch

Step right up. Best nickel cigars on the beach for a quarter

High striker

four-foot-long piece of strong oak. At one end of the lever was a section of automobile tire that kept breaking apart until I replaced it with a stubby rubber cylinder. The other end was part of an automobile spring that could be adjusted for the height reached by the missile up the vertical track.

Operating the high striker involved skills nicely possessed by my buddy Paul Ryan. He was a natural crowd pleaser, barking "Let's go, everybody plays!" The players get three hits of the hammer for twenty-five cents. Each strike of the bell gives them a cigar. I soon emulated my buddy and we'd shout "Everybody plays—best nickel cigars on the beach for a quarter," that each cost me four cents wholesale. Or for humor we'd yell, "Ring the bell four out of three times and win a new Cadillac automobile," or, "Three

Paul Ryan in later years

tries for a quarter, four tries for fifty cents—go for four tries, bargain price!" The operation was a moneymaker. Operating it was a fun experience. My income for the first July Fourth, was sixty-five dollars. The cost of building the device was fifty-five dollars. So I made my investment back on the first night, which was with a bit of guilt knowing that my dad and his partners never recouped their investment. My high striker provided a fair source of income for three summers. It was at the time that singer Tony Bennett charmed the country with "Because of You." Hearing that song is remindful of my adventures back then. When I was later drafted into the Army I gave the high striker to Paul, who soon donated it to his church. Revere Beach is no longer a seaside resort but an assemblage of condominiums, with no hint of its vibrant past.

In 1948 I got my Social Security card and began a life of conventional employment. My first job was as a factory worker. It was at Ross Shoe Company, in Lynn, a summer job that blistered my hands and was hard work. My task was busting finished shoes from their wooden lasts—a tiring task that blistered my hands. At least I learned something interesting, that the average shoe size for men was 8½ (today it's 10½).

To put myself on a path consistent with my talent for art, I phased into painting signs, which added a small bit to earnings. In those days there were no allowances from parents. Every penny was earned. Mowing lawns and picking berries to sell in summer, clearing snow in winter, and my newspaper route and sign painting provided an adequate income in my early teen years. My passion for boxing was not lucrative, fifteen dollars per bout, since most of my boxing activity was for sports events and "smokers" where a hat was passed around.

After graduating from high school in 1949 I became an errand boy for Vose Swain Engraving Company, near South Station in downtown Boston. This occurred when I was boxing. Instead of taking the underground train to deliver packages, I did "road work" and ran the distances. My minimum wage of sixty cents per hour wasn't bad, for each and every minute I was ahead by a penny while on the job. A pleasant adjunct was that my good friend Paul Ryan worked in Sinclair Valentine Ink Company, a block away from my workplace, which meant we could enjoy lunches together. A memorable lunch was on a pier next to a bridge that spanned one of the many inlets of Boston Harbor. He bet me a dollar that he could cross the bridge hand by hand atop the bottom flange of the I-beam that spanned the water. When he was halfway across I was aghast to see him losing his grip and thought he was joking. He wasn't, and fell to the water below. He was soaking wet, our lunch hour was cut short, and we hurried to his job site where he could dry. (He didn't pay me the dollar. Many decades later, before he died, he instructed his wife Norma to pay me that dollar!)

Something I learned while being an errand boy was witnessing the interiors of many workplaces and the disparity in different environments. In some, the employees seemed content and appreciated doing a good day's work, while in other very similar shops employees seemed oppressed and unhappy. Did the workers know of this? Did a worker in an oppressive environment know that a similar company down the street would provide a much more pleasurable work environment? Having access to the mailrooms of so many businesses gave me a valuable view that I'd miss otherwise. When my minimum wage was increased to sixty-five cents per hour, the company laid-off one errand boy, which was me. I enjoyed this work while it lasted, for it gave me firsthand opportunity to experience life in Boston, a city I was fond of.

I worked for a year in a chemical bottling plant until my dad found something closer to my field of interest—a temporary job in art and design. The gas utility company in Malden needed a window designer helper to replace a person on medical leave for three months. I jumped at the opportunity, which entailed a small battery of tests. Departing from company policy, a company official told my dad that I got a perfect score in mechanical ability. When my dad reported this to me, we both had a good laugh, for I was not the least bit interested in being a mechanic for trucks and cars. Mechanical ability was completely misunderstood by my dad and me.

On Memorial Day in 1950 my pal Peter Theologou arranged for us to take a couple of girls from Lynn for a ride in my spiffy chartreuse convertible. Peter drove and pretty Lorraine Barnes sat in the front seat and later changed to sit with me in the back seat. At last I had a potential girlfriend. I liked her a lot but was quite sure that she didn't like me as much. How could I get her to like me? I asked this of my window display co-workers who answered directly: Go beyond hugging and kissing.

Me and my first love

Wow! Taking their advice, with Lorraine's consent, my co-workers were correct. She became my first girlfriend. I was more than infatuated with her. Lorraine was my first love. All the love songs of the day had a fuller meaning. And nicely, both parents seemed quite okay with our romance. All this was wonderful. I couldn't have been happier.

Lorraine had previously told me that she had been in a sort of girlfriend-boyfriend relationship earlier with a boy named Dorsey who had joined the

Army. At summer's end I visited her home unannounced, opened the door and found her closely sitting on the couch with a uniformed soldier. She was with Dorsey! Instead of excusing myself and leaving, figuring I'd return when Dorsey went back to military service, I instead opted for the worst-case scenario, which was that Lorraine loved him and not me. I allowed myself to become devastated. The stupidity of youth knows no bounds. With brainlessness coupled with my infantile behavior, our romance was finished.

Just before my three-month employment at the gas company expired, I saw how posters were silkscreened. How intriguing that a squeegee could push ink through a silk mesh to produce art. Although I watched this process for only a few minutes, in my mind I could become a silkscreen printer for further employment.

Answering an ad in the newspaper for silkscreen printers, I was hired at Norman Industries in Boston where I actually learned silkscreen printing. I also met Ernie Brown, who I mentioned at the end of the previous chapter. Ernie was one of two stencil cutters who sat at art tables and artfully produced the films that were adhered to the silk screens. Ernie made his living *sitting*, not standing on his feet for eight hours a day. Stencil cutting would become my way of making a living. But my time at Norman Industries was short lived. Ernie told me that he was anxious to return to Florida and invited me and others to join him after Christmas. My parents approved this joint venture and my dad helped me put my Pontiac convertible up on blocks in our home driveway. We covered the spotlight, the chrome horns on the vender, and all vulnerable parts of the car. With a broken heart due to my mishap with Lorraine I was ready for a change.

So at the end of 1950 Ernie and I quit Norman Industries as soon as we received our pay and Christmas bonus, and headed south to Miami, where Ernie had visited before. Others declined this adventure. Driving south in his new Ford I wanted to hug the first palm tree that came into view. When spotting one, Ernie pulled over and I hugged this symbol of friendly warm weather—much in contrast to the bitter cold left behind. Upon reaching Miami we were lucky to gain employment before our meager funds ran out. I got a busboy job at the famed cafeteria Wolfie's in Miami Beach. Soon after, I was fortunate to find employment at Florart, a small silk screening company across the street from which Ernie found employment where my stencil cutting and silk-screening skills were honed. Summer heat in Florida, before the advent of air conditioning, became difficult, and being homesick for my family in Saugus, I thanked Florart for the several months of employment and returned north to my family just before July 1951.

My dad was acquainted with Mr. Tomajian, the owner of Revere Glass Company, who needed some silk screening done. I visited him to find he had a government contract to produce safety mirrors for emergency kits in life rafts. The idea was that stranded people in a life raft could signal their presence to aircraft by flashing signals of reflected sunlight. Directions for use were to be silk screened on the back of the mirrors. Thinking this was a short-term consulting job to get a printing operation in motion, I told him my fee would be five dollars per hour. This was when a silkscreen printer worked for less than one dollar per hour. Tomajian was furious, and said my fee was too much. I countered that he'd get his money's worth, for I could produce the screens and get his operation up and going. Reluctantly, he agreed. To my surprise, I worked forty hours per week for several months—at five dollars per hour—a great misunderstanding in my favor. Furthermore, he hired my brother Dave for one dollar per hour, and we enjoyed working side by side. This proved helpful to Dave, who continued in silk-screen printing on electrical circuit boards and more at Data General, where he worked for most of his professional life.

Sometimes it's the little things that make a big difference. I noted that when Tomajian was on the telephone with a supplier, before ending the call he'd ask, "Can I depend on that?" This made a world of difference, giving him a fuller picture of the other person's intent. Thanks to Tomajian I've used the same expression many times—often getting greater accuracy in agreements. In any event, Tomajian did get his money's worth with the silkscreened mirrors, Dave got a career path, and I earned enough to purchase a two-year-old black Buick sedan. In those times a person's income was earned, not distributed. Life was good.

As Tomajian employment was running its course, I rented part of a space in an industrial painting factory in Malden where I started my own business—Northeastern Silk Screening Company. I was already doing small jobs of screening identification information copy on newly painted machines for the owner. My hopes were to screen T-shirts as I had done at Florart in Miami. This was not to be, for a "police action" that turned into warfare was occurring in faraway Korea.

I never had a problem with passing a physical exam for boxing. Doctors would check the heart and all, and give me an okay. In 1951, when I was twenty years of age, the Korean conflict was on full blast and I got my induction notice for military service while I was in Florida. At the physical, I was deemed too light for my height (5 foot 10.5 inches and 110 pounds). I flunked the physical and was not drafted. However, I was called upon for a repeat physical a year later. This occurred when I was in Massachusetts. This time in 1953 was different. The examining doctor accused me of dieting to avoid being drafted. He stamped "passed" on the documents and I was drafted into the US Army.

FOUR

The Army Years

On a Saturday afternoon in May 1953, a jeep arrived at my home in Saugus to drive me to the induction center. I told my mom to have beans and franks ready that night, for when the military doctors discovered my low weight, they'd send me home. I wasn't sent home but instead was transformed from being an American citizen to an American soldier—I was Government Issue, abbreviated GI. To my surprise I was never weighed again until halfway

Beyond boy scouting
–the U.S. Army

through basic training at Camp Atterbury in Evansville, Indiana. At that point the army doctor said I could opt out of military service if I wished. I wrote to my dad, and he suggested I continue and serve my country. The camaraderie with soldiers was high, and in spite of the hardships of basic training, I enjoyed and benefited from the army experience and didn't opt out. As good luck would have it, a truce with North Korea was signed when my basic training was completed. So there was no combat, which I would have dreaded. I couldn't hunt animals, let alone humans.

A couple of weeks before being drafted I bought a second-hand guitar and had visions of being a country singer, for I admired Les Paul for his guitar playing. Or I could be a guitarist in an orchestra. I loved guitar music and took lessons for a couple of weeks from a local guitar teacher. I brought my guitar to Camp Atterbury and was so interested in it that I ordered a new custom Gibson left-handed cutaway for the astounding price of $175. It was a lot of money, but I saved a lot while working for Revere Glass Company. It would take several months for delivery of the guitar. This gave me ample time in the barracks to hone my skills. But alas, I discovered my skills were nil, which I learned by noting how other beginning guitar players progressed significantly faster than I did. Was this because I started left-handed when I should have started right-handed? After all, I've always been left-handed when writing, drawing, painting, so why not for playing a guitar. I box and throw a ball right-handed, and kick with my right foot. When I clap, I do so as a right-hander. Whatever the reason, my lack of progress signaled no professional future involving a guitar. When the Gibson guitar arrived at my home in Saugus, I picked it up on my first one-week leave. But it served mainly as an item to hock at pawnshops, for I and other GIs were notorious for running out of money before the monthly payday. I have the guitar today, displayed proudly with other memorabilia on my bookcase.

An important lesson learned as a new soldier was the importance of hierarchy between the ranks of enlisted men. Arriving at Camp Atterbury a couple of days before basic training, I chatted with a corporal who was later to be our trainer. When basic combat training commenced, he had an entirely different personality—more authoritarian. On a march he said to straighten up. In a good mood, I responded to him as "balloon ass." Oops, that was out of bounds. I was written up and given a so-so punishment. It was a lesson learned. Social discourse between a soldier and his officer is without humor. A fighting unit without that difference in rank would not be effective in combat. It can potentially undermine the chain of command, order, and discipline. Being a soldier is serious business.

Midway through basic training we were given the opportunity of a weekend leave off base. To attain this brief time to explore Indianapolis we had to pass an inspection of our barracks. We scrubbed the floors, the walls, and even the ceiling. The latrine was sparkling clean with all brass polished, and the windows with their sills nicely cleaned. Upon finishing, our lieutenant entered the barracks—wearing white gloves. We shuddered. He pulled a footlocker to the middle of the room and stood on it to reach a heating duct. He opened a door to the duct and swiped his hand inside. We had not expected he'd inspect so remote a place. Showing dust on his glove, he explained that the barracks were filthy. We failed the inspection. Being a couple of years older than the others, I thought the scenario was predictably a part of training to vex our emotions—and a bit amusing. I almost smiled as others in the barracks were dismayed. That was a part of army life. The next week we passed inspection and got our first chance off base.

Before induction into the Army I had met Gloria at the Oceanview Ballroom. She was from Newton, Massachusetts, the well-to-do part of the Boston suburbs. I asked for her telephone number and was lucky enough to get a date with her. At a nightclub in Newburyport, we saw a live performance of The Ink Spots, a favorite singing group in its day. When I drove her home, I was gentlemanly enough not to try kissing her goodnight. In basic training everyone was lonely for girlfriends or potential girlfriends back home, including me. I was lonely for Gloria, who I hardly knew. We exchanged letters. When a leave was granted, I drove with buddies back to Massachusetts. At the beginning of the trip, I noticed a slight itch on my lip. Oh no, not the dreaded cold sores! It took two days to drive home, and by the time I reached Newton my lip was an enormous and ugly cold sore. The curse that messed up New Year's Eves at the roller-skating rink was in full force. Gloria was not sympathetic. She was already making a great concession dating a skinny working-class guy, and she didn't even ask me into her home. It was over between us.

There was a fellow in our barracks who wrote letters for his inarticulate buddies. He was a gifted writer and boosted the romances of more than one guy in the barracks. His literary efforts for one particular guy went too far, the rest of us thought. His love letters led to a marriage at the end of basic training. Wow! We could only guess how long that marriage would last.

On a visit to Chicago while at Camp Atterbury in 1953, I purchased a green Buick convertible. Several of us enjoyed getting away from GI-overrun Indianapolis (Naptown), and driving north to Peoria, and to Chicago, where we frequented jazz clubs. I saw Charlie Parker, a world-famous American jazz saxophonist play there, something that I enjoy mentioning to jazz enthusiasts. On one trip back from the north with some buddies, I made a wrong turn and the Buick skidded out of control. No one was hurt, but the car

My green convertible later painted fire-engine red

banged a few trees before stopping in a cornfield. My buddies and I were close to being killed. Very scary. We survived without serious injury and only slight car damage. We continued our drive back to camp. While the car was being repaired, I had it painted a fire-engine red, which was more appealing to fellow soldiers at the base who rented it. With the money from short-time car rentals I was able to pay off the car's loan.

My art put to good use

I found a niche for my sign painting skills in the army, which began as being the guy who lettered the numbers one through sixteen on the block of wood fixed to the pole carrying our company flag. For each of the sixteen weeks I'd paint over the previous number and increase it by one. I established myself as a sign painter. This led to lettering desk signs for the officers and areas of the supply room, painting murals in the mess hall, and so on. In short, I was essential to the battalion, a part of the Dixie Division. So whereas others were sent abroad and to other locations after basic training, I was retained in the Dixie Division. As soldiers, we were referred to as "Darling Daddys."

The Army wished to integrate black soldiers into our unit. At first it was awkward because they overused the derogatory word "motherfucker" that was annoying to most of the other guys. A quite shocking word. But a comedian might humorously consider the term as it refers to one's father—the one who *is* intimate with one's mother. As time went on, for most of us, it became barracks chatter. Another degrading cuss term was "cocksucker," which again could simply refer to a man's wife or girlfriend. Comedian George Carlin years later referred to such terms in this way. Cultural adjustments were made, and all went well with diversity in our barracks.

In 1954, military officials closed down Camp Atterbury and the whole camp caravanned cross country to Camp Carson in Colorado Springs in Colorado. We slept in tents along the way. In Colorado we became affiliated with the Fifth Army, regarded as a prestigious step upward from the Darling Daddys of the Dixie Division (as we were called in the local bars in Indiana). I wondered about girls back in Naptown with their previously fashionable darling-daddy tattoos, now that no Dixie Division remained. I enjoyed the several-day caravan since army life was more relaxed than it was in the sixteen weeks of basic training.

My official duty in the Fifth Army was being part of a team in an artillery battalion of 105-millimeter howitzers, popular cannons that were common in Korea. It was interesting to stand directly in back of one to see the shell exiting the barrel when fired. It appeared as a disk that quickly shrunk to a point as it traveled away from the cannon. Although my official duty was being on the howitzer team, my actual assignment was sprucing up the battalion offices and buildings with good signs and murals. A humorous incident occurred with many onlookers when I was painting signs in the supply room atop a scaffold. When I finished, I held up my brush and asked the guys in an innocent sounding voice, "Where should I stick this brush?" The predictable answer came as a loud chorus. I was a sort of entertainer as well as a sign painter.

The best thing about being in the army was the camaraderie with buddies from all parts of the country. Only one soldier, Sid Chapman, was from my Saugus hometown. I could understand why my boxing mentor Eddie McCarthy had earlier reenlisted when the Korean War was in full swing. He reenlisted on the condition that he rejoin his missed buddies in Korea.

Being part of a group, all experiencing a common situation, is the appeal of military service. That appeal, unfortunately, led to Eddie's death in Korea.

A downside of military service is the lonely separation from wives, girl-friends, or other loved ones—for those who have them. Soldiers craved being with girls. Girls were always on their mind. Even when marching in cadence we sang "I know a girl right over the hill—she won't do it, but her sister will. Sound off, sound off, one two three four, sound off!" We did the same when I became part of the battalion's elite Drill Team.

When camping on bivouac, girls were especially on our minds. Seeing grassy ground beneath bushes, or in any private place, led to imagining how it would be with a girl there. During a brief break while marching on a very hot day, I asked my buddies to imagine off to the left of the road Marilyn Monroe lying naked in a bed waiting for a GI (everybody was in lust with Marilyn then). On the other side of the road was a table with generous slices of ice-cold watermelon, waiting to be slurped and eaten. Which side would my buddies choose? Only if you were there would you know the answer was the watermelon. Lust takes a back seat when you're thirsty and craving a cold drink. With a few exceptions, all soldiers were starved for sex and were girl crazy. For most GIs, intimacy with girls were rare. We would have gone nuts if women soldiers were in our midst, which was unconceivable at the time.

Camp Carson had a weekly newspaper. Once my duties had been established, I visited the newspaper office and arranged to draw a comic strip "Private Chooch," for publication. I continued the strip until my tour of duty

ended. I was gratified to progress from the many comic strips I created as a young teen to the professional quality comics in the Army. I signed all my work *Hewitt Drew it.*

I also made a few bucks in the barracks by selling "chances." I used the dayroom typewriter to type a list of names of prospective girls twice on each line on a sheet of paper. When I cut the paper into twenty strips, I'd sell each strip for one dollar. Half of the strip would go into a military helmet, and the other half to the buyer. When all strips were sold, someone would draw the winning half strip from the helmet. The winner got ten dollars on the spot. I also got ten dollars for thinking of the idea. The guys were okay with that—especially those who won.

Being in the Army meant always being tired. Up before sunup and on the go all day long. We were always busy and lacked sleep. More than once in midday I crawled under my bed in the barracks and slept for a long nap. I was lucky that I was never caught. And nights out on the town were stretched as late as possible, guaranteeing a tired next day. If the Army made men out of boys, it made tired men.

The scenery in Colorado Springs was beautiful, especially close to Cheyenne Mountain (which later was hollowed out to become the national defense center). I fell in love with Colorado Springs and the nearby mountains. The soil was a reddish color not seen in New England, and the redness was highlighted at the Garden of the Gods nearby Manitou Springs. Some memories are engraved in my mind. A vivid one was listening to Rosemary Clooney singing "Hey There" on my car radio while driving through Manitou Springs. My favorite song of all time, "I'm Getting Sentimental Over You," by Tommy Dorsey was also memorable. Although I was quite infatuated with Colorado, my barracks buddies weren't. They were homesick for their hometowns, mostly in Massachusetts and upstate New York. Much of their free time was spent longing for their familiar home surroundings and listening to audiotaped music played by Eastern disc jockeys that were updated each time they'd get a leave home. This only intensified their homesickness. In my mind, they were blowing it by not discovering what beautiful Colorado had to offer. There is more to appreciate than home in this world.

Another vivid memory is leaving the barracks in the early morning to secure items in the supply room. Although the radio announced it was zero degrees outside, I was comfortable wearing only a T-shirt. Zero degrees in the Boston area meant cold cold, demanding a heavy coat. The dryness of Colorado air was pleasant, both in winter and in summer. Hot days in summer were much more comfortable than same-temperature days on the East Coast. Colorado has seasons without the brutality of freezing weather in winter and soaking-wet hot in summer.

I was a couple of years older than most of my fellow soldiers and in better physical condition due to my boxing years. I came in second in our battalion physical fitness rankings. I excelled in the forced marches of basic training. Also, I didn't require the amount of water that my buddies required. So in breaks they were grateful that I shared my water with them. Those marches *were* brutal. Soldiers who couldn't continue were picked up by medics and driven back to the base. To succumb to that was very much looked down upon, and very few soldiers faked a ride back.

The fellow that ranked first in physical fitness was a guy I'll just call Luis, a short, stocky weightlifter from Puerto Rico. Luis was a born-again Christian who was quite outspoken about his devotion to God. He told me that he wanted to be forgiven for his airport activities in New York City after he emigrated from Puerto Rico. He'd meet strangers as they departed from planes and welcome them to the United States. It seemed wonderful to newly arrived immigrants, apprehensive about their new country, to be greeted by a representative who spoke their own language. Luis helped by carrying their bags—but then before they realized it, literally stealing them. Luis robbed his new countrymen and left them destitute in their new country, something that later caused him considerable guilt. He indeed needed God's forgiveness.

Thanks to former years in boxer training, I performed well in marches, carrying the same heavy load as guys twice my weight earned me a lot of respect from my fellow soldiers. I also felt good as a member of my battalion's

That's me in Row six

drill team (pictured). I was also good with a rifle. I ranked first in marksman-
ship in my battalion with M-1 rifles and received a cigarette lighter with our
battalion emblem, which I sent to my dad. My parents and almost everyone
I knew smoked at the time. I was never a smoker, attributing that to an anti-
smoking lecture years earlier by an elderly intoxicated man on the streets of
Lynn. His reasoning made sense, and I heard him. (I've always been prone to
take the advice of others who seemed to know more about a thing than I did,
which later helped in writing physics textbooks.) But my main reason for not
smoking had to do with boxing. Losing your breath in the ring is a no-no.

I thought the months in the Army would result in my gaining weight,
as it did for many other skinny guys. It didn't. Another my Saugus pal Sid
Chapman was another skinny guy, who became a lifelong friend and joined
me in activities after discharge. We seemed immune to weight change, while
many overweight guys lost weight. My resistance to gaining weight wasn't
that I didn't eat, for my appetite was good, and although everybody complained about the food, I found it quite good. I was a slow eater as evidenced
by being first in the chow line, but the last to leave the chow hall. This was
because I chewed my food thoroughly. When I was a child, my mom sat
beside me when I ate and made sure I chewed my food multiple times before

swallowing it all. This likely aided good health, though I seemed cursed to be skinny always. To this day, I'm still a slow eater and chew my food multiple times before swallowing.

That I was a "bone sack" was pointed out by Frank Shelby, with whom I played guitar. Another friend was Scadova, almost as skinny as me, and anything but handsome. Frank, on the other hand was a lookalike of actor Paul Newman. In walking through parks in Denver, it was something to see how girls gazed at Frank as we three walked by. No girls ever looked at me and Scadova like that. Frank was so fine looking and confident of himself that on one occasion he met a girl who was to be married that very weekend. His confident and flirty way won the girl's heart, and he had his way with her all night long in her truck. Scadova and I slept in my Buick. Oh, to be Frank Shelby for just one weekend! And to add insult to injury, Frank unashamedly told me and Scadova that no girl would ever be attracted to us—we weren't good looking enough. He said we should just accept it as a fact of life. And I did. But my admiration of Frank was diminished. As with my school experience in the third grade when Mrs. Beckman held my skinny arms up for the class to see, you don't tell people of personal deficiencies that they sure as hell already know about!

Most of my buddies in the army drank a lot when off base. I joined them in drinking, but I was a light drinker, mainly due to the influence of my father. My dad couldn't *hold* his drinks. If he had one beer, or his favorite, ale, he'd have to have another. Then another. It was most unpleasant to see his friend Gene Jenkins and other drinking buddies drag him up the front stairs of our home after an afternoon at the bars, his feet loosely thumping along the steps. Or worse, Gene carrying him over his shoulder like a sack of potatoes. As luck would have it, dad was a *good* drunk, and not a *mean* one. He'd pull me on his lap and with bulging eyes hug me and tell me how much he loved me, while scraping my face with his lightly shaved whiskers, along with the detestable odor of alcohol. So my orientation toward drinking wasn't the joyful experience most of my buddies had. Getting drunk was rare for me.

Having one or two beers when playing music was another story. Frank Shelby, Scadova, and I took our guitars to clubs and beer halls in Pueblo, about forty miles south of Camp Carson. Places with "open mikes" would welcome patrons who wished to sing and play for drinks. Frank was a skilled guitarist and a good singer while Scadova and I simply strummed chords to accompany him. At a particular crowded beer hall, I took on my identity as Huey Paul. After a few beers I sang "Release Me," which was one of the popular songs of the day, changing the words a bit for humor. That proved to be another blow to any idea that I could make it as a singer. No drinks from the crowd then, or ever, as I remember. But we did have fun.

I mostly enjoyed playing guitar with a new friend who had just returned from Korea and joined our battalion. That was Eddie Vandenburg from Oklahoma. Eddie wasn't reluctant to tell war stories. He told of how in Korea with a recoilless rifle he picked off more than one Chinese soldier on the other side of a wide valley, for China had sent troops to fight on the side of North Korea. With telescopic sights he could see his adversaries clearly,

Eddie and Jean Vandenburg

and he delighted in shooting them. (Were any of them relatives of my present wife Lillian?) Eddie had a lovely bride, Jean. We had good times together, and Eddie and Jean, along with Sid Chapman, are the only military friends I've kept in contact with over the decades.

Any homosexuality in the army went unnoticed for the most part. But in the first days of basic training a very feminine fellow, after an interview with a psychologist, was soon cleaning out his locker. We never saw him again. Toward the end of our enlistment, however, a large burly guy, who I'll just call Bill, was extra chummy with Al, the mailroom guy. Al was funny, and whenever someone would exclaim "Jesus Christ!" Al would retort, "Jesus Christ, is *he* here too?" This got a laugh from most of the guys, including me. At that time, I took my Christianity casually and couldn't imagine a God that lacked a sense of humor. Bill and Al frequently slept in the same bed. Nothing was said of it. For one thing, Bill was a tough guy, and nobody wanted to offend him. If homosexuality was a problem in the army, it wasn't evident in our outfit—with one exception. An older top sergeant who was in charge of the mess hall was caught inappropriately touching a younger-looking GI, a rosy-cheeked boy—the kind that gays were likely attracted to. Although the sergeant had a silver star for outstanding service in combat, and served for many years with a spotless record, he was discharged from the army. We saw that as unjust and very sad.

In our outfit was a bully, whom I'll call Jim. What Jim enjoyed was seeking gay guys in bars, leading them on, and then beating them up. Jim was sadistic and enjoyed battering people. I had run-ins with him myself. One time we almost came to blows. I asked him to settle things outside with boxing gloves. Although Jim was a strong muscular guy, I would have had no trouble with him—as long as gloves were kept on. Gloves are awkward for a street fighter, but home-sweet-home to a boxer. In another altercation, we again almost got into a fight without boxing gloves. I persuaded him that if he beat me up, he'd have to answer to all the guys about fighting with the skinniest guy in the barracks. But if I beat him up, he'd have the humiliation of being beaten by this skinny guy. I explained that either way it was a lose-lose situation for him—and to buzz off. Looking back, I think I would have lost any fight with Jim on his terms. To this day I don't know, for I've never had a fist fight with anyone since I learned to box. All my fighting experience had been by the rules of the gentlemanly art of self-defense.

The last I heard of psychopathic Jim was that after his army discharge, he joined the police force in Lynn, Massachusetts. I always wonder if he got his wish to shoot somebody in the forehead, something he made no secret of. Whether or not he got that wish, it's easy to imagine many unfortunates that must have been Jim's victims—all, supposedly, "within the law."

One of my army activities was giving reenlistment talks, for I thought that people who lacked passions or skills for civilian life were assured employment and a position in the military (assuming peacetime activity!). I exempted myself, for I was going to be a comic strip artist, maybe a short story writer, and secure a solid future. I attained the rank of a corporal in short order. And about the time I was discharged in May 1955, I was promised a promotion to sergeant if I reenlisted. But I had greater plans—to escape the working class by discovering uranium and becoming rich. This was at a time of great general interest in uranium prospecting. Upon being discharged from the army, I remained in Colorado to prospect for uranium.

My interest in uranium was fueled by an article in *Look* magazine that I found in the day room with a cover photo of a gruffly guy, Vernon Pick, with a caption about how he made $9 million in one afternoon. Vernon had sold a uranium claim he staked in Colorado. I reasoned that if Vernon Pick made that much money in one afternoon, I'd spend *many* afternoons in pursuit of uranium, especially since I was in Colorado. Of course I did realize that the "one afternoon" was the culmination of many months of prospecting. Anyhow, I stepped up the rental of my red Buick convertible to fellow GIs and saved enough money to purchase a Geiger counter. So off I went on weekends into the nearby Colorado mountains in search of uranium.

Before being discharged from the army, my buddies Del Enman and Richard Acquaviva and I discovered the scenic mountain town of Salida, about a hundred miles west of Colorado Springs. Its population was about 7000 and it was nicely located in the middle of the Rocky Mountains. It was a railroad center appropriately declared as "Salida—the Heart of the Rockies." It was also a center for rock hounds, with its ample minerals and old abandoned mines. In our first Salida visit we walked into the Rocket Bar. How welcoming that the owner realized we were soldiers and gave us a round of drinks on the house—a very different situation than soldiers experienced in the crowded bars of Colorado Springs. We had discovered a special place. That was the beginning of my love affair with Salida, which would be my base of prospecting activities.

Downtown Salida   The meeting place

The Rocket Bar and a few other bars in the middle of town were friendly gathering places. Soon we discovered the small dance hall called the Country Club on the edge of town that became our favorite. It was a large single-story hall with a potbelly stove on one side and a stage in front for live music. Saxophone, piano, and drums by Chris Trujillo and his wife Julia and son Andy provided wonderful dancing music. I brought my tape recorder to the club and recorded many of their songs. A new local friend, Pal Linza, introduced us to potential dancing partners.

A couple of weeks later at the Country Club dance hall I saw a lovely girl who is permanently etched in my mind—Millie Luna. She was standing at the entrance with a cigarette dangling from her lipstick-red lips, wearing her brother's untucked white shirt. She watched as I walked from my table to approach her. I asked if she'd like to dance with me. She looked me up and down and simply said "No." Later I learned that the girl next to her, Shirley, remarked that I must have tuberculosis or something, being so skinny. But more important, Millie didn't know that the red Buick convertible in the parking lot belonged to me! Aha, she underestimated her potential suitor. I ignored her the following week to teach her a lesson. When we finally danced together, we hit it off splendidly. She introduced herself as Millie Lunea, from Paris. I kidded her back saying I was some celebrity heartthrob of the time. We enjoyed each other, neither of us having a clue that we would one day become husband and wife.

Being a soldier was good for me, for it was without combat. As I mentioned, a great part of army life is the camaraderie with fellow soldiers. My initial provincial views were greatly widened by living side by side with guys who had different cultural backgrounds from different parts of the country. I benefited greatly by this. I saw tough guys from big cities blend in with people they wouldn't normally encounter. Underlying this was conformance with army rules, without which there would be chaos. Getting along with one another was more than impressive. In combat it was essential. Grudges between soldiers were to be settled, for in combat they'd be lethal. An unfair commander hated by his troops, for example, would have more than the enemy to fear. We'd all heard "friendly-fire" stories galore. Adhering to army rules was enhanced by stories of being sent to the stockade where sadists awaited newcomers, and the horror of being wire brushed in showers. It was a matter of both self and group interest to get along. "Yes sir, no sir, no excuse sir" were the permitted choices of response to superiors. We had to learn to "like it." We were one group and saluted one flag. Solidarity was the result. We all got along.

These characteristics of soldiering are why I favor a draft of all citizens at the age of eighteen into national service. I was drafted into the army and am glad of it, especially since I didn't face combat. A draft need not mean military service. National service should be broader than that. Think of the nonmilitary programs before World War II such as the Civilian Conservation Corps, or the achievements of the Peace Corps, which offer great personal possibilities while making great social contributions. In a nationwide setting, all citizens would be a part of a single "army"— national camaraderie big time! Pledging allegiance to the same flag may seem squirmy for some, especially with the newly added "under God," that was influenced by evangelist Billy Graham and added a year before my tour of duty ended. Graham's influence defined the times. A few weeks after my discharge, he and like-minded others persuaded President Dwight D. Eisenhower to insert "In God We Trust" on all our currency. My thinking was that religion or lack of it is a personal thing, and shouldn't be political. This huge focus on religion was *then*, and need not be now. Focus would instead be respect for common-sense national rules—all nonsectarian. Draftees who cannot offer that respect would be welcome to emigrate from the USA to any country whose flag and rules they'd better appreciate. Say what you will of America, no citizens are prevented from leaving—instead, vast numbers of people are clamoring to enter. To me, that's a biggie. Hence I think the USA would benefit by eighteen or twenty-four months of national service from all its citizens. Downsides? Yes. A big one is a disruption of young people's lives. But I see a much greater upside, mainly national cohesiveness.

I have always felt this way, and fast-forwarding to today's disunity. I see national service as all the more important.

FIVE

Uranium Prospecting and Other Adventures

M y two years in the army were mostly spent in Colorado Springs, a few years before the famed Air Force Academy came to be there. I was lucky that the white flag was raised in Korea at the end of basic training. No more fighting. Whew! I never experienced combat. My duties were mainly painting signs and

creating a weekly comic strip in the camp news-paper, even though I was officially in an artillery battalion. At about the time of my discharge, the base was renamed Fort Carson. Instead of return-ing to my hometown in Saugus, Massachusetts, I elected to remain in

My sign shop and prospecting headquarters in Salida

My treasured jeep and scintillation counter

Colorado in hopes of striking it rich with uranium. As it happened, what funds I earned in Colorado were not from finding uranium, but from painting signs in the town of Salida.

Salida was a lovely town, and I took to it right away. I rented a small cabin, a shack really, for seven dollars a week on an alley behind West Sackett Street in the middle of town. It had a kitchen, a bathtub, and a potbelly iron stove in the main room for heating. It was just right for my sign painting and prospecting activities. I became the town sign painter, always busy with interesting work for interesting people, which provided a source of funds for my prospecting activities. I drove to Denver and traded my Buick for a compact four-wheel drive Willys Jeep station wagon. I painted a nice "Hewitt Drew it!" logo on its doors. The jeep served well for hauling sign-painting equipment to sites around Salida, and afforded sleeping quarters when I and others were prospecting. Its four-wheel drive enabled it to travel on roads that an ordinary car would never access. I loved my Jeep.

Soon my brother Dave and friend Ernie Brown came from Massachusetts to join the prospecting adventure. Ernie was very helpful in

Welcome Ernie Brown

designing many of my signs, as well as trudging along rock outcroppings in search of uranium. In our three-bed cozy cabin we played music via a 45-rpm record player with two choices of music. One was the album "Music for Lovers Only," by Jackie Gleason, and the other was a single, "Time Goes By," by Marty Robbins. We played both records over and over, setting a nice tone. More than a year later I attended a Marty Robbins concert. When Marty asked for requests, I raised my hand and yelled "Time Goes By." I was heartbroken when he replied that he didn't recall that song. Back in our cabin, live

Some music with our Gardunio neighbors

music was enjoyed with our neighbors, the Gardunio twin brothers. Guitar music set a festive mood.

One afternoon while driving from nearby Buena Vista to Salida I picked up a hitchhiker. His name was Marion Zimmerman. He was a scruffy guy, into his early sixties or more and handicapped with a deformed left arm. I asked how far he was going. He replied as far as I was going and to drop him off any place where he could hitch a new ride. It began to rain. I noticed he had a Bible with lots of red underlines and notes. I wanted to help this good Christian man and said he should stay at my place for the night and continue when the weather was favorable. Marion stayed at my humble shack for more than a decade. I was quite okay with that. Interestingly, I later learned that his Bible scribbles were places he saw as erroneous. He had contempt for religion. Oh well, different strokes for different folks.

Painting signs put food on the table. Another source of income was tied to something I found interesting. Submarine sandwiches were common in Massachusetts, but quite unknown in Colorado. Aha! I'd start a sandwich business, a family business with Millie and her two sisters and introduce cold submarine sandwiches to Salida. I arranged with a local bakery to supply sub-sized buns each morning and bought cold cuts at a wholesale meat distributor. A sub contained several types of meats, cheese, lettuce, onion, a bit of olive oil, and a small hot pepper at the end of a bun. The subs were wrapped in cellophane and sold for fifty cents each. I created a cardboard box to contain a dozen or so sandwiches, nicely lettered, "Paul's meal in a bun—50¢," and a cartoon of a guy eating one. We'd deliver boxes of a dozen or so to local bars and a couple of stores and place the box near the checkout counters. At the end of each day we'd pick up the leftovers and make thirty-five cents for each sandwich sold. But business wasn't good, mainly due to the useless leftovers, and it was made worse by putting in less meat and more lettuce. Uh-oh, this came to a head when a customer loudly stated that she didn't want to spend fifty cents for a bun full of lettuce. So our sandwich business gradually declined. At that time I remember being fascinated with tacos, which were unknown back East. If I were in Massachusetts instead of Colorado, I'd introduce tacos to East-Coast people. In 1962 a fellow named Glen Bell founded the first Taco Bell eatery—which became much more lucrative than Paul's meal in a bun.

My prized possession

Climbing over mountains armed with both a Geiger counter and a scintillation counter was not an easy task. Many hopeful prospectors who streamed into Colorado and Utah soon found the effort too much and left after a couple of months. Prospecting was hard and the chances of success bleak. Dave and Ernie also left after about six months, which was quite understandable.

A couple of prospectors, Frank Winton from Oklahoma and Jack Bennett from Michigan, stopped by my shack and asked to borrow my scintillation counter. We all agreed with the provision that if they found uranium that I get an equal share. After a couple of weeks they returned, reporting some good readings in a national forest in the Sangre de Cristo Mountains, some sixty miles south of Salida. The radioactive findings were some two miles up Cotton Creek Canyon that was fair game for prospectors. The mountain range was on the edge of the huge San Luis Valley. Together, we returned for further exploration and found an outcropping that extended from one side of the valley to the other. Alpha or beta readings were low, but steady. We enlisted a geologist to investigate our findings and made a report. Apparently we had found a layer of uraninite that spanned the valley—a lot, but of low grade. The hottest rock we found assayed at 0.16% uranium. Commercial ore, by contrast, at its lowest grade is assayed at 0.20%. Not very promising.

Having zero knowledge of physics at the time, I was puzzled that clouds seemed to perpetually hang over the mountains, but not over the valley. But why? Years later this would be an intriguing physics topic.

Winter snows were coming and the area would soon be inaccessible. So it would be off to Massachusetts for me, and back to Oklahoma for Frank,

With Sid Chapman at Cotton Creek

and to Michigan for Jack. That was after we staked the claims, made maps, completed the required forms to secure the claims recorded in the assessor's office in Saguache County. Frank told me something shocking about Jack—that he didn't believe in God. Uh-oh, anybody that doesn't believe in the creator of nature cannot be trusted. I had never met someone who does not believe in God. As it happened, Jack wasn't confident of the success of our uranium venture in our part of Colorado and dropped out of the picture, which was a welcomed relief. Frank and I, however, remained convinced of success with the claims. Our adventure was soon joined by Sid Chapman, my Saugus buddy from the Army.

It was comforting to know that when the uranium claims were sold, I'd be a wealthy man. What would I do with my riches? One of the first things I'd do would be to pay off the home mortgages of neighbors on Fairview Avenue in Saugus. When I mentioned this to my mother, she was quick to teach me yet another of life's lessons. If I did that, she said, I would no longer "fit in." Despite my generosity, I'd be resented by those neighbors who would see me as lifting myself above them. My mom went on to say that if I became rich, I'd have to move to a location where other rich people lived—where I'd fit in. Hmm. This gave me much to think about.

Another prized possession: my 1953 Cadillac convertible

With a confidence of commercial success, I traded my Jeep for a two-year old 1953 apple green Cadillac convertible in Denver—a stunningly beautiful automobile. The payments were a hefty one hundred dollars per month, but affordable, especially with the prospect that the following summer we'd be affluent, if not rich, from the uranium findings.

With snow cutting us off from the uranium claims, Millie accompanied me on the drive to Massachusetts—the reason being that she was pregnant! Uh-oh . . . the girlfriends of my friends became pregnant, leading to quick weddings—*shotgun weddings,* as they were called. Despite my love of Millie, at this stage of life I certainly wasn't ready for marriage. Millie felt the same and we kept our mishap a secret. She would have the baby in a home for unwed mothers on the outskirts of Boston, which arranged adoption to parents who desired a newborn. Having a baby out of wedlock in the 1950s was viewed as shameful and deplorable. With great shame, an unmarried mother and her family would be socially disgraced. Many girls face an unexpected pregnancy would "leave town" and secretly give her baby up for adoption, very different today. We returned to Salida the following summer.

Fast-forward to 2014. The baby that Millie tearfully gave away to adoption in 1956 resurfaced, but sadly after Millie's death. Jean Marie Hurrell came into the Hewitt family after she researched her birth mother's records. This was wonderful! How much more wonderful it would have been if Millie

Yoga instructor
Marie with dad
Philip Hurrell

First meeting with daughter Jean

Jean with daughters
Marie and Kara Mae Hurrell

lived long enough to experience the love of Jean, who luckily had a very good upbringing, first in Saugus, then in neighboring Wakefield, and finally in Wilmington—where the town cemetery contains the remains of her ancestors and mine on my mother's side of the family. We love Jean and her family. A very happy outcome!

To make a long story short, nothing ever came of the uranium claims we staked. Financially, the adventure was truly pie in the sky and was the focus of summer activities for some four years. (Fast-forward some fifty years later when I wished to show the claims to my rock-hound nephew-in-law, Bill Candler. When we drove to the mouth of Cotton Creek, we were shocked to find that the road I so frequently used in trucking mining equipment was now covered with fast growing mature aspen trees! No trees were present in that road in the 1950s. We could not travel the two-mile trail to our former campsite. Things change over time.)

Two of my sign customers in Salida were very impressed with the large signs I designed and built for them and encouraged me to attend college, and to give thought to becoming an engineer. I contemplated the idea, but not much. One concern was how would I eat during college years? About this time, Ernie Brown returned to Salida and convinced me to return to Miami with him for winter employment. Once in Miami, I was lucky to be hired by Webster Outdoor Advertising Company to paint billboards. I was initially assigned to a construction crew that erected the steel frames that support billboards, usually atop buildings.

A sign for my best customer

I was happy to be employed. I became friends with a co-worker, Frank, one of about five black guys that worked at Webster. There was no racial animosity between us. None, even in the deep South. One night Frank introduced me to some of Miami's nightlife in his part of town. With Frank, I was quite comfortable being the only white person in the bar. Frank reminded me that if the case were reversed, he'd feel quite uncomfortable. At Webster, we were paid on Friday afternoons. Shockingly, there were two lines for receiving pay envelopes—one line for about fifteen whites and the other for the five blacks. This was the South in 1956. Egads!

At this time I missed my Colorado sweetheart, Millie Luna. On the telephone I proposed marriage and sent a letter asking the same of her mother. Her mother answered yes, but asked if I'd considered that I was entering into an interracial marriage. Millie was Mexican and I was Caucasian. I loved her and the interracial issue was not a problem for Millie or for me. We were both raised in households to accept others for who they are whatever their different origins. My mother's assertion that most people are decent certainly applied to Millie and her family. However, a bigger consideration was that Millie and her two younger sisters developed rheumatic fever in childhood that led to serious inflammatory disease of the valves of the heart and can shorten one's lifespan. I accepted that Millie might not live past fifty. (She lived to sixty-eight.) Soon she arrived by train (flying was expensive and unpopular in those days).

On December 22, 1956, we had a small wedding at the White Temple Methodist Church in Miami that was attended by Ernie Brown and a parishioner witness. Our honeymoon was spent at the Allison Hotel in Miami Beach, and on the advice of local friends I checked with the hotel beforehand to be assured they would allow interracial honeymoon couples. We were accepted. However, on our wedding night a Caucasian honeymooning couple in the hotel nightclub was acknowledged on microphone by the hotel host while we were ignored. We had no problem with that, and simply recognized that we were in the southern culture—something we accepted. When Millie was a child in Salida, she similarly tolerated the policy of "no Mexicans" in the hot-springs community swimming pool.

Honeymoon

Wedding witnessed
by Ernie Brown

Our honeymoon photo

We're all dealt a deck of cards when we're born. Fairness in the deck is a matter of luck and we play that deck the best we can, without complaint. So do your best with whatever you are dealt. That was the lesson that Millie's mom Ruth Luna gave to all her children. She also said that in troubled times to just "let it be." I came to appreciate this years later when the Beatles better defined *letting it be*.

A later photo of Burl Grey

At Webster I progressed from constructing signs to painting them. I was assigned to paint with a fellow that others preferred not to work with. This was Burl Grey, a highly decorated war veteran who was part of the Normandy invasion of World War II. Fellow painters didn't like working with Burl because he was odd, too intellectual, and didn't talk of sports, girls, and cars the way that most guys did. Instead he talked of ideas—not relished by other painters. Burl was glad to find I was open-minded and welcomed him opening my doors to thinking. I was intrigued each day with his war stories, but mainly with his intellectual puzzles and ideas about science. Eight hours per day with him was stimulating with a capital S. Each evening I'd share much of what we talked about with Millie—things I'd never thought about. Burl was outspoken in telling me that there were nonbiblical reasons for the sky being blue, clouds being white, and people interacting with one another in the ways they did. When I protested that people were basically decent, Burl responded by saying that if toilet paper were taken away from them for a week they'd become monsters. Burl was a behaviorist who saw human behavior as stimulus-response, all shaped by our environment—that we were little more than machines. Needless to say, except for the machine bit and other excesses, I found many of his ideas challenging and enormously interesting.

Jacque Fresco

One of Fresco's models of the future

Burl introduced me to Jacque Fresco, a futurist who espoused much of what Burl had discussed with me day after day for months. Millie and I found Jacque captivating, and we signed up for his small weekly classes at his studio home in nearby Coral Gables. The emphasis of his lectures was that science and engineering, all human oriented, would best lead to a better and kinder world. He had many stories that applied logic to everyday situations. He spoke of a fellow who gave five-dollar bills to people passing by him every Friday afternoon at a particular time. During the fourth week he showed up with no money to give. What was the response of his would-be recipients? They beat him up. Is there a lesson here? He told of many similar stories, all with a moral perspective. In more important and larger scenarios he described how the engineering of sensibly designed cities could elevate human wellbeing.

I asked Jacque about his belief in God. His reply was ambiguous, that if there were no God, the nature we see around us would be no different. He felt the hypothesis of a personal God that watches over us was extraneous—one with no confirming evidence, and suggested I read *Man and His Gods* by Homer Smith. The book had a foreword written by Albert Einstein. Also, to read Bertrand Russell's *Why I Am Not a Christian*. Was I betrayed at the age of fourteen when in answer to my question about God's existence my minister didn't tell me that Einstein's religious beliefs were akin to those of the philosopher Spinoza? And that Einstein's spelling of God was N-A-T-U-R-E? Jacque expanded my thinking. If science had better answers for what's happening in our surroundings than ancient biblical accounts, I should look into science. Millie and I had never been impressed with a more enthralling and influential individual as Jacque Fresco. Burl and Jacque changed the course of our lives.

Newly married, Millie and I had an ordinary and satisfactory sex life. Fresco added some gusto to that. He was an advocate of free love, inspired by Bertrand Russell's book, *Marriage and Morals*. The gist was that all living creatures, including humans, have sexual urges that ensure survival of the species. These urges, however, may involve more than one's mate. If instances of sexual involvement with a nonmate occurs, such needn't violate the bond between married couples. With beforehand agreement between husband and wife, any such extra-marital encounters could be accepted, and not the prevalent grounds for divorce. This would be very difficult for those prone to jealousy or possessiveness. Some couples could accept this, some couldn't. Being alerted to the dangers to a sustained marriage, Millie and I were both acceptors. Such adventures of freedom were infrequent and a private thing with us over the years that we valued and never regretted.

On December 3, 1957, a few days after our son Paul was born in Salida, I turned twenty-six. We packed our belongings and left again for Massachusetts to start a new life. My perception that college takes too long a time was addressed by my sister's husband, John Suchocki. When I told him I would be thirty-two years old if and when I graduated from college, he slowly and deliberately asked me how old would I be at that time if I didn't go to college. That did it! I prepared for a college education. Sometimes the simplest of ideas doesn't catch on until somebody better articulates them. John went further saying that physics would show how mathematics underlies all mechanical things. As an example, he explained the math

Our first son Paul

of a balanced seesaw with people of different weights—how the light guy sit a longer distance from the pivot, the fulcrum, to balance a heavier partner. It was all numbers—beautiful mathematics. I'd have a good time learning these things. Maybe I should say relearning, for all this may have been covered in my ninth-grade Basic Math for Boys class. If it was, it didn't have the same clout. We learn when we're ready for it—when information is valued. I could hardly wait for a new beginning to my education.

Committing myself to such a huge swerve in life didn't make sense to my fellow workers in the silk-screen printing industry. They spoke to me of the financial disadvantages of such a move, enumerating the wages lost in the college years and the many years to recoup them. Their reasons didn't affect my decision because college was not in quest of better finances, but in quest of a meaningful life in science, even though I didn't know what that might be.

I was fortunate that the GI Bill provided a stipend for my education, which I could further supplement by part-time work in silk screen printing firms. I want to jump ahead a bit and just relate a finding that I found amazing—the workplace of the college educated. As a physics student I joined an afternoon field trip to the nearby Avco Research and Development facility. The building was modern with fluorescent lighting everywhere, a big parking lot for employees' cars, shiny floors, and cleanliness wherever I looked— employees in white shirts with ties, water coolers, coffee pots, great desks to work at. The workers seemed interested in their work. No one had to "look busy" and no one went to the toilet frequently to rest legs from standing too long. What a contrast to the screen shops of Boston with dirt and grime everywhere, poor lighting, in a gloomy atmosphere. If my friends at the shops could see how education makes a huge difference in how one spends their work life, they would better understand why I left for school.

In education I entered a much improved and exciting phase of my life.

SIX

Student Years—Lowell Technological Institute

Spending two years in the US Army provided me with the GI Bill, a route to college that included the working class. I didn't take advantage of the bill until I was inspired by friends to take college seriously. Foremost was Jacque Fresco in Miami who pointed the way to a life in science. So in late 1957, with my wife and infant son, I drove from Colorado to Massachusetts, where I ventured into the admissions office of the Massachusetts Institute of Technology (MIT). A few minutes into an interview with an admissions official, another official was summoned to join the discussion.

Before them was a naïve twenty-six-year-old married man with an infant, with scarcely a high-school science background, wanting to become a scientist to contribute to a better world. He had no funds but had prospective uranium holdings in Colorado that might provide needed finances in addition to the GI Bill. He was a most unusual candidate. The first official said to the other, in my presence, that sometimes an institution such as theirs ought to, from time to time, take a gamble on someone outside the usual box—that enthusiasm and seriousness of purpose may make up for other deficiencies. Being shy, I said nothing. Afterward I realized that I should have told them that *I* was such a person and would bring credit to being that exception. I blew it by saying nothing, which likely indicated that despite my enthusiasm, I lacked determination, confidence, and smarts. The interviewers suggested I attend preparatory school to make up for high-school deficiencies and then attend the more affordable Lowell Technological Institute (LTI)—the state supported science and engineering college which is now the University of Massachusetts Lowell. I took their advice.

I enrolled at Newman Preparatory School on Newbury Street in Boston for three semesters of algebra, geometry, trigonometry, and physics. I excelled in all my classes. My teachers were top notch. I learned from them. Mr. Jim McDonald was the physics teacher, and I loved his course. He made physics interesting and taught us to view physics and other courses in science like stacking bricks in a vertical column. Each brick supports the one above it. Likewise, each physics concept supports the one above it, connecting to the next, and the next. This logical thinking got me hooked. Physics would be my direction.

Then something quite momentous occurred in my studies. I discovered the answer to an unsolved question that puzzled Burl Grey and me on our Miami sign-painting scaffolds. When painting together we pondered

the relative amounts of tension in the pair of ropes that supported our scaffold. How did the tensions change with variations in our distances from them? We were stymied. Back then we had no rule to guide our thinking. When I discovered the *equilibrium rule*, an offshoot of Newton's first law of motion, I shared this discovery in a letter to Burl. Aha, for a scaffold at rest, all the supporting forces acting on it balance to zero. So the sum of the forces

My guide was always "Why?"

up must be canceled by the sum of the forces down. How simple! Nature obeys its rules. I was happy to be studying them and was in love with physics!

During my stint at Newman Prep my dad found living quarters for Millie, baby son Paul, and me in Winthrop, a small city close to Boston. We busied ourselves painting the walls of our new home. On a wall above my place of study I hung a small picture frame with the short question, "Why?" The question was to keep me focused on the excessive study I was to undergo—an effort that would lead to an understanding of physics and the math associated with it. This sign would follow me to a later home in Lowell, and then years later to Utah. If my field of study were engineering, it would have been more appropriate that the sign read "How?"

National interest in physics and engineering was spurred by the launching of Sputnik, the first artificial Earth satellite, by Russia. My interest in science was pre-Sputnik, sparked by Burl Grey and Jacque Fresco. I was already taking classes at Newman Prep when Millie and I, just after sunset, would find delight in watching Sputnik moving across the sky. Although we were in the dark, the orbiting satellite was still being bathed by sunlight and could be clearly viewed by observers below. Who at the time could have imagined that this curious Sputnik would be followed by swarms of orbiting satellites that would pinpoint one's location on Earth's surface in the GPS of smartphones—a feat to dwarf the aftermath of the first printing press. I was happy to be a bit ahead of the curve when the country progressed into the space age.

While at Newman Prep I worked part time as a silkscreen printer where I was pleased to experience a connection between the physics I was learning and my everyday environment. At the printing shop was a guy in charge of paint supplies. Printers came to him with buckets that needed refills. He'd use a large spoon to fill the buckets by dipping into big drums of paint. Sometimes more than one printer would wait for his bucket to be filled. Seeing a solution to this slow process I suggested a different tool: Instead of a spoon, why not use a larger-capacity soup ladle, the utensil that cooks use to dish soups from pots to bowls. Looking at me quizzically, the paint guy responded, "But I've always used a spoon." I thought, aha, Newton's first law of inertia: A body at rest tends to remain at rest. A guy accustomed to a spoon will continue to use a spoon. Physics truly is everywhere.

When I wasn't cutting stencils, I was on my feet pushing a squeegee for hours on end—tiring work. I found it puzzling that a geometry teacher at Newman Prep mildly complained about the hard work of teaching. My thoughts were that this fellow never worked in a factory. It seemed to me that any exertion associated with teaching would be mild in comparison.

Lowell Technological Institute

In the fall of 1958, after three semesters at Newman Prep, I entered Lowell Technological Institute, about twenty-eight miles north of my hometown in Saugus. To augment the GI Bill payment ($160 frustratingly sporadic monthly checks) I supplemented this amount with earnings from my stencil cutting for silkscreen firms in the Boston area. We luckily rented a cold-water flat near the school that was nicely affordable, seven dollars a week—but it had no heating system and no hot water. Heat was supplied in winter by keeping the oven door open in the kitchen and water was heated as needed on the gas stove. One of the first items I acquired was a used chalkboard from my dad's office to mount on a bedroom wall. I knew that learning was best achieved by discussions and interactions with other students. I'd invite physics buddies from my classes to weekend and evening chalkboard sessions. One was Robert Hulsman, about my age; through our study sessions we were able to keep up with our younger counterparts. Like me, Robert was married with a son. He was still active in the US Air Force and eventually became an engineering professor at New Mexico State University. He was the only one in our study group to earn a PhD. Our friendship continues to this day.

During the first semester in Lowell my second child was born, November 16, 1958. Her birth occurred less than a year after son Paul was born in Colorado. I first saw my daughter in the hospital baby ward, hoping any of the other better-looking babies was mine. Little Leslie Ann Hewitt was jaundiced with tiny slit eyes. When the nurse picked her up and held her against the glass window and asked me if she was beautiful, she was aghast when I replied "no." The nurse said that she couldn't believe that a father didn't see his baby as beautiful. I said that I was just being objective, that although she was mine, she in fact was sickly-looking and quite a bit less than beautiful. That was the budding scientist in me. Fortunately, little Leslie's looks improved dramatically during the following days, with open

My pretty Leslie

eyes and a full head of dark hair. From my teens, I kept a diary from time to time. On November 28 diary entry, "Little Leslie is doing fine, is cute although she was very ugly at first." From then on, Leslie was always in the front row when it came to looks, and in addition, she had the charm that often comes with good looks. Leslie has been my source of pride ever since.

Because of the required heavy course load at Lowell Tech, I had difficulty keeping up. Being ten years older than my fellow students didn't help. All physics majors took the same courses in a four-year program with seventy units of physics and thirty-four units of math. This was more physics and math than most universities require for a PhD in physics. I called it "information overload," a condition I've combatted ever since. Several of the textbooks we used were graduate-school textbooks at MIT. LTI was trying to out-compete MIT, but on paper only—it lacked the highly competent professors of MIT.

An interesting course at LTI was electronics. Although talk of the newly invented transistor was in the air, our focus was on vacuum tubes. At that time all radios and television sets operated via vacuum tubes. Sensing they would soon be replaced by transistors, we nevertheless mastered the niceties of vacuum tubes and completed our course. As important, computers were making headway and we learned programming via Fortran, which involved punch cards. With better computer facilities at MIT, we traveled there almost weekly to run our programs. Who then imagined the coming age of integrated circuits and computers that were soon to follow.

My school experience at LTI was wracked with financial problems and academic insecurity, which wasn't helped when Richard Ivers, the dean of students, on opening day told us to look to the person on the right, then on the left—that one of them wouldn't make it to graduation day. I feared that *I* was that unfortunate person. If a glance in some magic crystal ball had assured me that I *would* graduate on time in 1962, my school experience would have been less stressful. Of twenty-one physics majors that made it to the senior year, I ranked fourth, with a large gap between me and the top three students—Jack Kennealy, Ray Marcotte, and Peter Panousis. Despite that good showing, my grade point average for the four years was only 2.4, a C, which limited my options for graduate school.

Many of my financial and emotional difficulties were recorded in my diary. On January 1, 1962. "Present plans: Almost five years of endless studies now—due for a change. For weeks now been anxious to get my studying and work over with, and get out of school to something more enjoyable and relaxing—want to be able to visit friends on a weeknight if I care to, or even sit with my hon (wife Millie) and watch TV without knowing studies are being ignored—want to have the time to read what I like, even time to think and do things—these cannot really be done while studying at college—not with relaxation—tired of having to hit books every single night. At present I have applied to several grad schools, as I want the advanced degrees that apparently are so necessary for a career in teaching, a career which besides being of greatest interest to me, would afford the "time" to live which I so miss now. If the present should be remembered in the future as a time when more time was available than seems to be the case, it must be remembered, any time taken for any activity is done so with the shadow of demanding studies always in view—herein lies the big factor—so don't know whether to go on in grad school majoring in physics, or take an easier course, education. Ideal, I think, would be an MS in physics, and PhD in education. I'm thinking now for the first time in any serious vein of high school teaching, at least for a breather in my goal of teaching at the college level. Anxious to cash in a few chips from these long years of study—never spent so long a time on any one thing—want to go west or south." I endured this frustration in both undergraduate and graduate school.

Most of the stress of school was financial. Making ends meet was a continual challenge. My folks helped as much as they could, and relatives would loan us ten dollars from time to time. Friend Ernie Brown and other friends seemed to be forever bailing us out with loans. Loans via Household Finance were frequent. Wife Millie, after much searching, got a night-time job in an electronics firm that lasted a few months, which greatly helped. Another point of stress was that Millie might get pregnant again. Her greatest fear was missing her period. Even getting condoms wasn't always easy. One time we pulled our car in front of a pharmacy, and I went in to replenish our supply. The pharmacy was a husband-and-wife operation. While the husband was busy with a customer, and with the car waiting outside, I approached his wife and asked for a dozen Trojans. The husband overheard my request and came running across the floor yelling how dare I ask his wife for such! I was asked to leave and not return. Birth control was shameful back then, and condoms were sold under the counter. This prevailing attitude didn't change until the "sexual revolution" of the late sixties.

Also stressful was the time spent in securing and delivering freelance stencil cutting jobs. My skills at cutting stencils for silkscreen printing provided needed funds, but the efforts of pickups and deliveries often took more time than the actual work. Then, to make matters worse, the Cadillac convertible became troublesome. In cold weather it wouldn't start, the battery and its replacements didn't hold their charge, and its bald tires seemed to be always going flat. I was lucky that my dad was helpful with car repairs.

I experienced a different social life at LTI than most students because I was among the very small group of married students. During our second year in Lowell, we were accepted into public housing. This was wonderful! Rent was thirty-five dollars a month for a two-story three-bedroom home, including heat and electricity, and sidewalks cleared of snow when needed. We happily painted the entire interior and built beautiful furniture and made

Public housing like this

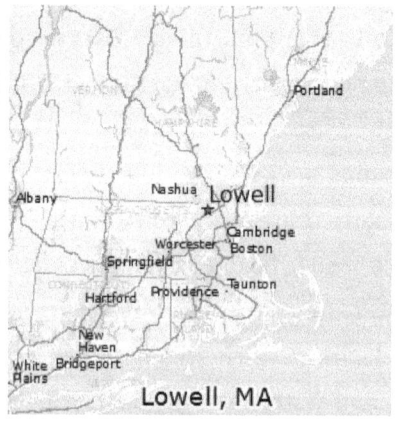

Lowell, MA

drapes. We'd invite fellow students to home-cooked meals. Sometimes we'd be invited to their homes for the same. Our social life while living in the projects was much more than satisfactory.

Some people in public housing seemed to be in a different "class" than the few married students lucky enough to live there. In chatting with a young man who lived a few doors down the street I let him know that he could borrow my typewriter. He came for it when we were away for the weekend, and left a note, "I came to procure your typewriter." Whoops! Why "procure?" I guess he was writing to his "superior," a college student who presumably spoke a more refined language. I thought this perceived personal difference in social class was unfortunate.

Living in public housing was living in heaven. However, it seemed strange to us that many neighbors didn't appreciate what we loved. We later realized that how much something is valued is greatly influenced by what was had beforehand. Spending the previous year in our cold-water flat greatly helped to appreciate the many amenities of living in a housing project. Perhaps our disgruntled neighbors always lived in public housing and took for granted its conveniences. For example, some neighbors left windows open to control the heat rather than adjust the thermostat. We shoveled our own steps after snowstorms while neighbors complained of the slow service of city workers clearing the snow. Even the free trash bin was often not used when discarding trash. Rather than pulling the entry trash door down, residents would simply toss their trash at the foot of the trash facility. Not all neighbors were so crude. But there were enough to give thought to our liberal views.

On March 19, 1960, during my sophomore year, the phone rang in the kitchen and when I answered it my mother told me that my dad had just died. I was shocked. I asked for no explanations. I simply told her we'd be there within the hour. Fortunately, friend Ernie Brown was visiting and agreed to stay with the kids while we tearfully drove to Saugus, some twenty-eight miles south. My dad was only fifty-two and died while fixing a leak under the kitchen sink. It wasn't determined whether he had a stroke or a heart attack. What was clear was that he was gone. He was loved by hundreds of people who showed up at the funeral. The funeral director, Bisbee, said that he hadn't had that many people at a funeral before. My dad was known for his generosity, integrity, and friendly manner. People felt good being with him. A strange and surprising thing occurred at the funeral. Greeting the large numbers of people, those I loved and those I hardly knew, was an uplifting experience. I sensed guilt that I was enjoying the company of so many people at a time I was devastated by the loss of my dad. But that's what funeral parlors are for, to pro-

My dad

vide social activity that diminishes the sense of loss. Some relatives said that my father's death would be more than I could bear, and I'd not succeed in school. But I knew my father was cheering me on, as he had at my boxing bouts when I could hear his voice clearly above the crowd. I strived all the harder at Lowell Tech.

We were lucky that family was there for us. To complement a life of study, social balance was provided mainly with brother Dave and his wife Barbara,

Dave, Barbara with Nancy, Mom, Margie with kids,
me, son Paul, John Suchocki with young John, and Millie

who were our best friends. Holiday gatherings were mostly with sister Marjorie and her husband John. Aunts on both sides of the family were also central to our social life. On my dad's side was Aunt Dot and her daughter Jeannie, and on my mom's side was Aunt Ruth and Aunt Minnie. I very much valued the emotional and social support of family during my college years in Massachusetts.

Charlie's grown-up kids:
Mike, Danny, Irene, Carol, Cheryl

We spent summers in Salida to be with relatives, mainly Millie's mom Ruth. And to alleviate an ongoing financial problem, some bucks could be made painting signs. We enjoyed visiting Millie's older brother Charlie and his wife Lucie, in nearby Los Alamos, New Mexico. Charlie had been a bartender during the atomic bomb era and I relished his stories about physicists that he served drinks to who worked there. We also enjoyed the five Luna kids.

At Lowell Tech I discovered my skill for writing after I turned in my first lab report and was called into the dean's office. The instructor and dean thought that my write-up was "too-well written," and must have been copied from elsewhere. I explained that I simply wrote the way I talked. My writing ability was enhanced by contributions I made to the college newspaper, *The Text*. I was voted editor-in-chief in my senior year, and I put a lot of effort into transforming the newspaper and enlisting a cadre of new writers. On controversial issues we presented articles for both sides of an argument. The writers of most articles had their photograph with their columns, acknowledging their contributions. One columnist was Howard Brand who became a high school physics teacher and remains a close and lifelong friend. The coverage of articles and columns was well balanced, making *The Text* a highly respected newspaper. Little did I know I was setting the groundwork for a career as an author that was to follow.

Science editor Howard Brand and editor-in-chief

Although I have never written a "letter to the editor" to a newspaper or magazine, I wrote one to *Time* magazine. It was published. My letter favored humanism, an alternate to conventional religions. I was surprised to get a visit by Harold Rafton, a humanist devotee from Lowell. Harold found where I lived to personally congratulate me. From his expensive car and demeanor he seemed obviously wealthy, and here he was—in the housing project with a young obviously poor man with a wife and two children. Sometimes wealthy people give financial help to those in need. It crossed my mind that Harold might just do that. I read his expressions and figured that he thought it would be a bit demeaning to offer a financial gift—he didn't wish to interrupt the hardship that was building my character. My expression said he was wrong—that some extra bucks would be enormously appreciated—my biggest problems were keeping bills paid and food on the table. But

I kept quiet. Expressions were not matched by words. He patted us on our heads, thanked me for writing the letter, and departed. I never saw him again.

Before any notions of textbook writing, I felt that my future would be in teaching. Inspired some years earlier by Jacque Fresco in Miami, I saw that none of the professors at LTI could hold a candle to his teaching ability. I believed that with more physics knowledge I could do a better job teaching than I witnessed in the classrooms. As an October 11, 1960, diary entry states: "I'm disgusted to sit in classroom all day and put up with men who can't teach—I want to teach, myself, so badly!" With an eye to future

ANDREW A. OUELLETTE
B.S.
Professor,
Physics and Mathematics

Best instructor

teaching, I came to view my classroom experience as lessons about what *not* to do, more than what to do. On the positive side, I learned a lot from Mr. Andrew Ouellette, who taught math courses. Every day we got down to business—beginning with clean and clear solutions to homework. Then he'd progress to an overview of new material, skillfully matching the textbook coverage, and concluding with one or two sample problems. His simple routine was very much valued and we learned a lot. Mr. Ouellette, was voted the best instructor at LTI in a survey published in *The Text*.

On the negative side there was Mr. Epstein, who challenged the patience of many of us to remain in physics. His lectures were clear and reinforced by his industry experience. He knew his subject and delivered it well. But his exams and grading system were heartbreaking. Instead of presenting problems akin to those we'd studied in class and in homework, his problems, were remote from core material. I suppose he had the attitude of wanting to see who in the class could manage challenging material, whether or not it pertained to his lessons. But worse, the grades he assigned to top scorers in the class were usually about 50 percent, with the rest of us scoring in the "noise level." At the end of the course he had to curve the grades, or flunk most of the class. His assessment practices were extremely frustrating. Studying was not rewarded. Because of my disdain for his practices, and unfortunately of him, I learned very little from Mr. Epstein. To make the situation worse, he taught the core physics courses, so it wasn't possible to avoid his classes.

From this painful experience I vowed to do two major things when it became my turn at teaching. (1) To assure that my exams covered a question or problem for each major part of the material covered. (2) Assuming my top students merited an A, to create exams wherein top students would score 100 percent or very close to it. For a large class, the grading distribution should approximate a bell-shaped curve, ranging from below 50 percent to 100 percent. Students would then not be subjected to the whims of grade curving at the end of the course. These two policies ensured a fair course—very important!

Another teaching policy was learned the hard way for me—exams must be new. An instructor should not take the lazy route and use previous exams. The only course I flunked at LTI was electricity and magnetism taught by Professor Louis Block, whose final exam was identical to the one he used the previous year. I was unaware of that since I lived off campus and the exam was given on a snowy morning. Getting to school was problematic and I arrived about five minutes late for the exam. I pored myself into it, which was not only difficult, but also puzzling because classmates all around me were finishing it early. This was most unusual and I was the last to finish—and the only one to fail the course. Professor Block explained to me in his office that since my score was appreciably lower than the scores of my classmates, and my previous scores were nothing to write home about, he had the responsibility of failing me. I didn't have the courage to tell Professor Block that the higher scores of others were due to a before-class study session of his last year's final exam that included everyone but me. The *F* grade certainly didn't help my grade point average. But repeating the course the following semester helped me better understand electricity and magnetism. Being stronger in that area or subject of physics certainly was of value. Also important was the humbling experience of failing a course, something all teachers should experience firsthand. And more important, it taught me to go through the effort of preparing new exams each semester so that no student or group of students would have an unfair advantage. Again, fairness is essential to a well-taught class.

In a seminar course in public relations, a quirk of mine came to light—not being competitive. Our professor stated that he found it remarkable how throughout the course I could at any point have taken command of the student group—but didn't. He wondered why. I didn't respond, but today I know that I was glad to be part of the group with no need to be number one in the group. I've always been a bit bothered with people's quest to "get ahead." Get ahead of whom? My focus has been to perform to the best of my abilities. I say *my* abilities, not those of others. The US Army recruiting slogan had it right: "Be All You Can Be," rather than being better than the others. I'm not a competitive person, which later sometimes was amusing to friends who saw my no-need-to-win composure in playing dominoes. I valued playing more than winning.

In 1962 I graduated from LTI with a BS degree in physics, which was an enormous relief. My four years seemed to be an obstacle course. I can more accurately say that I got *through* LTI. Getting *into* physics would hopefully follow in graduate school—somewhere else, because I wished to step off what seemed like a treadmill experience. I needed a break badly. My wish was to

A yum day with Millie and kids

learn physics that I then only partially understood. I applied to eleven graduate schools, but with my C average and in need of a teaching assistantship, I was painfully rejected by all of them.

Millie and I preferred a western location and we had our hopes pinned on Colorado State University. I was accepted, but without the offer of an assistantship. With my GI benefits about to expire, and a wife and two children to feed, this wouldn't work out. Neighboring Utah State University (USU) in Logan, Utah, was a hoped-for possibility due to a connection between labs at Lowell Tech and USU. After some back and forth correspondence we were elated to be approved at USU, along with a teaching assistantship!

The Cadillac convertible continued having mechanical problems, so I sold it for a low price and bought a 1954 yellow four-door Plymouth, a more dependable car that got us from Massachusetts to Utah in 1962. It was a very nice family car, and I've preferred four-door cars ever since. I built a trailer and loaded it with our belongings for the long haul. After we settled into our student housing in Logan, I left the empty trailer in a nearby parking lot and was crushed to find it stolen on the second night. Apparently some students who were ending summer studies helped themselves to the trailer. Welcome to Utah.

SEVEN

Graduate Years—Utah State University

Beautiful Utah State University

U tah State University is a public land-grant research university in Logan, surrounded by the lovely mountains of Cache Valley about ninety miles north of Salt Lake City and a hundred miles south of Pocatello, Idaho. It's among the most beautiful campuses in the country—a nature paradise. We were provided pleasant student housing on campus and found people in Logan very friendly, on and off campus. We felt quite welcomed.

It's not trite to say that our most important possessions are our friends. On our drive from Massachusetts to Utah in 1962, I remember saying to Millie how there were people in Utah, then unknown to us, who would become our lifelong friends. Although our move was away from our current friends in Massachusetts, new friends awaited us. Sure enough, at a Unitarian Fellowship meeting at USU, we met newlyweds Huey and Sue Johnson. Huey, from Michigan, was an outdoorsman and we had summer picnics in nearby Logan Canyon and good times fishing at nearby Bear Lake. At their off-cam-

Huey Johnson with wife Sue

pus home in town we enjoyed barbequed clam burgers, a mixture of canned clams and hamburger, a favorite of Sue's.

Another friend was Ellen Drake, originally from Scott City, Kansas, who earned a master's degree in wildlife management. She had just returned from hitchhiking through India and Africa, and she wanted to aid wildlife management in Ethiopia.

Ellen Drake

Ellen in her research tower

When we commented on how courageous she was to hitchhike in foreign lands, she replied that she'd not do the same in America. Ellen's thesis was about studying rabbit behavior for an extended time from an observation hut perched above the rabbits. She had a similar project with African wildlife in Ethiopia after she graduated. The Johnsons and Ellen were our best friends in graduate school.

I was disappointed to find that I was nearly as stressed as a graduate student as I had been as an undergrad. Professor John Merrill announced that we'd have to do all the end-of-chapter problems in Goldstein's *Classical Mechanics*. Goldstein's textbook was very challenging, and solving all the chapter-end problems was close to a vertical uphill climb. Most of the problems were above my head, and solving them was a difficult task. I hadn't gotten into study groups with other students.

Huey was aware of my stress and came to my rescue. He presented a copy of John Steinbeck's *Tortilla Flat*, and requested that I read it—not later, but this weekend. No way, I said, for I was submerged in work that had to be done. Huey insisted. Now Huey has a compelling way about him that explains his success after graduate school. In addition to being especially articulate, he is blessed with an impressive deep, authoritative voice, and his suggestions are to be taken seriously.

That evening I began reading Steinbeck's book, which is about a central character, Danny, whose concerns in life were finding ways to get the next bottle of red wine for his friends and himself. No Goldstein's problems, no pushing beyond natural abilities—just going from one bottle of wine to the next—with a reverence for life in doing so. Reading the book was the psycho logical break I needed. That weekend Millie and I went down to the Cactus Club in downtown Logan, where drinking was legal, and Huey introduced us to Vino da Tavola, a good cheap wine. My stress level was lowered, my composure restored, and I got through the Goldstein problems okay and surmounted the academic hump. From then on, I addressed Huey as Dan. And in return, he referred to me as Dan. Two Dans, and as it happened, close and lifelong friends.

After Dan earned his master's degree in biology, he and Sue moved to San Francisco. He also convinced Millie and me that we also should settle in the same part of the country. San Francisco was politically liberal and was blessed with a mild climate with no snowy winters. Persuaded by Dan's advice, our destination became San Francisco. By the time we visited the Johnsons in Mill Valley, just north of San Francisco, Dan had been hired by The Nature Conservancy as its western regional director. Nothing beats starting at the top!

John Merrill

Farrell Edwards

At USU physics came alive! The physics professors, notably John Merrill and Farrell Edwards, both Mormon bishops and three-year-wonder PhD graduates in physics from Caltech, were inspirational. What a treat! All the professors I encountered at USU were excellent. I came to learn the physics that I struggled with at LTI. Much of this learning was due to being a recitation instructor with twice-a-week teaching sessions on problem solving. Students would attend a main lecture three times a week and divide into small groups twice a week where a graduate student would help them, mainly in problem solving. Being a recitation instructor was a giant step from my former chalkboard sessions in our Massachusetts home. Learning is accelerated when articulating the steps in problem solving. Teaching is an effective learning experience.

My first teaching assignment as a recitation instructor was not a smooth one, for I was terrified of standing before a classroom of engineering and science students, all strangers to me. My recitation sections were on Tuesdays and Thursdays, complementing the Monday-Wednesday-Friday lectures given by the professor. Opening day of the quarter was on a Tuesday, which meant that students would not have attended the first physics lecture by the professor. To me this meant no Tuesday recitation session. My first one would begin two days later, on Thursday. Oops! I was told I'd have to greet the students on that Tuesday and simply discuss vectors and general material. I was frightened. I refused. Fortunately, another graduate student took my place.

So my first meeting with students was on Thursday. I was quite prepared with carefully written notes. I remembered what Jacque Fresco taught in one of his home lectures in Miami—that if a shy person discovered fire in a theater, however shy, that person wouldn't hesitate to yell "Fire!" Why? Because the shy person delivered information that was *relevant* to those who heard the cry—certainly of value. If students viewed my lecture material as relevant, any shyness would be overcome. Hence my emphasis on notes to clarify the lecture material. Although well prepared, I was fearful and repeatedly went to the nearby men's room to urinate. Rather than enter the classroom a bit before ten minutes after the hour, I waited a couple of minutes more, not wanting to start talking while people were entering the room.

Finally I made my entrance and faced about thirty students. As I nervously introduced myself, I peed a little in my pants, like a nervous dog, enough to show on my light grey pants. Fortunately I was standing behind a lectern giving my lecture until my pants dried. My presentation finished about fifteen minutes shy of the expected fifty minutes, so I dismissed the class early. Subsequent recitation sessions went smoothly, for I always strived to be fully prepared with ample problem solutions. I soon found I was very popular with the students.

The program of courses at USU was fine, in line with earning a master's degree in one year. The one-year program aligned with my strong personal desire to accomplish this quickly. But there came a snag. Department chairman John Wood announced a change in my program. I'd have to take the graduate-level electricity-and-magnetism sequence to attain my degree. Since the sequence had already started, that meant I'd have to remain at USU for another full year. I was devastated. I reminded him that he had previously agreed that this wouldn't be necessary because I took the same course using the same textbook at LTI. He said that the graduate committee overlooked this requirement earlier and held firm. This was intolerable to me.

As luck would have it, my reputation for teaching reached Professor Jan Taylor, the chair of the school of education. She sat in on my recitation sessions and stated I should get the credentials necessary to teach high school. She assembled a program wherein I would receive a master's degree in secondary school teaching within my one-year time frame. I was already into a thesis project for a master's degree in physics, which would qualify for my education degree. So I relished this grand plan. I would complete my studies at USU in one year as initially planned.

During the following quarters I took education courses in psychology, testing methods, educational history, and reading courses to complete the program. Happily, some of the courses were requirements for the California teaching credential—a very nice embellishment. Furthermore, starved for funds, I was employed by the education department to evaluate secondary teachers' performances. I began with a prospective teacher who I thought did a satisfactory job of teaching, well deserving of a C. The C grade for a teacher candidate in the education department caused an uproar. To me, a C meant average with no serious deficiencies. I stuck to the C grade, which interestingly was then also supported by the education department. They knew grades were inflated in the education department. The following quarter I had two other teachers to monitor. Happily, one teacher was outstanding and received an A, and the other was very good and got a B. Two more candidates the following quarter were also better than average, and avoided a C. My belief that a C grade indicates average mastery of a subject remained throughout my succeeding years of teaching.

What kind of people become teachers? There are many, many for students choosing teaching as a career. My personal experience demonstrates one of the reasons. Quite simply, I couldn't attain a physics degree in my one year-slot, so rather than leave with no degree I switched to the department of education. Not complicated. Were many others like me—changing to lower academic hurdles? Why were education departments less demanding? In conversation with professors in education their response was simple to understand: Teacher shortages! State law required a warm body in a classroom to at least take student attendance. The education department couldn't be as selective as departments of physics, chemistry, and so on. Weeding out students was common in many fields, but seldom in education. When I came on board in the physics department there were two of us. The other fellow was called into the department chair John Wood's office and we never saw him again. Luckily, that was not me.

I was walking across campus on a Friday on November 22, 1963, when a student friend announced that President John F. Kennedy had just been assassinated. Oh no! This was enormously shocking news! Millie and I were devastated, for we were fans of the president. Classes at USU continued without pause, which was disturbing to me. This was not JFK country. Millie and I were glued to the TV the following evenings. The several speculations about this national tragedy added to our grief. And sadly, he was killed before he was able to know of the coming moon landing.

My life was on a favorable path but not without difficulties, both financial and emotional. I've kept a diary most of my life, and my entry for January 1, 1964, reports: "Hope to graduate in June with two degrees (MS in physics, MS in secondary school teaching). Then hope to start teaching physics in California in the fall. Outlook for past several years has been that when full-time teaching begins, *life will begin*! I'm so weary of doing homework, preparing for exams, taking subjects I am not interested in, but which are necessary toward attaining desired ends; spending an evening at the movies that's not at the expense of study time; the insecurity of attaining desired ends, the awareness of the richness life has to offer outside my physics office, and outside the student housing area, and outside of Utah; the anxiousness to *do* something constructive in line with public service. I have such a desire to be free of student responsibilities, tho realize I'll have greater ones later. Will welcome these. High hopes for 1964!!"

In addition to an MS in physics, my list in early 1964 included a PhD in education. At that time a master's degree was the route to teaching physics in a junior or community college, and a doctorate in physics was the route to doing physics research. A master's degree would be fine enough. I very much enjoyed my research activities at USU, which accounted for my two thesis projects in physics. The first was acceptable for the MS degree in education. The thesis project was devising and constructing a prototype spectrometer that could piggyback on a rocket and examine polarization characteristics of its exhaust plume. My device proved successful, even though it wasn't miniaturized for actual rocket travel.

Before completing the MS in secondary school teaching, I visited the San Francisco Bay Area to interview for a teaching position. I interviewed at Contra Costa College in San Pablo. Two weeks later I got a letter of rejection. I also interviewed at San Francisco State University, where I found encouragement, but only if I planned to continue graduate education to earn a doctorate in physics. I felt too academically exhausted for that. Then I visited City College of San Francisco (CCSF) and was interviewed by Dean Bill Mayo.

The CCSF interview was a disaster, for the dean broke my spirit and reduced me to the edge of tears. Dean Mayo had contempt for education degrees and asked what made me think I'd be worthy to teach at City College with a degree that wasn't worth the paper it was printed on. I also learned later he was contemptuous of Mormons, and since I was coming from Utah, he thought I was one. This was my first and only encounter with religious discrimination. When I returned to Utah I visited the home of Farrell Edwards, who urged that I continue with my studies in physics. Since my GI Bill had expired, to sweeten his suggestion he proposed I work for the Electrodynamics Laboratory that was associated with the physics department. A new thesis project would be in tandem with working at the lab. I agreed and returned to the physics department for another good year of learning, and very importantly, of employment. As it happened, I passed the qualification exam for the PhD, but I understood that attaining that degree would mean at least three more years at USU. Oops, my quirk for timing again—I couldn't delay permanent employment for so long a time. An MS in physics would be quite satisfactory—and more than that, a dream come true.

The MS thesis for physics consisted of laboratory measurements of spectra produced in rarefied helium gas that was bombarded with 100 keV protons. An additional parameter was measuring the polarization of light that made up the spectral lines, something similar to the task in my first thesis. For this research, the apparatus was already functioning, "just waiting for my measurements." My knack for writing was helpful in writing both theses. Although I enjoyed physics research, I enjoyed the prospect of teaching. Research is more *thing* oriented—teaching is more *people* oriented. I preferred the latter. To teach physics at a community college required an MS in physics, which was all I needed. If perchance I wanted to someday teach physics teachers, then an advanced degree might be needed, which is why the PhD in education was mentioned in my wish list of early 1964.

I learned a lot from frequent discussions with my USU physics classmates, who were all going on to PhDs in physics. Among them was my office mate, Robert Luke, along with Jay Phippen, Gary Layton, and Gordon Prather. I remember one topic we discussed was the idea of "arm-waving" when lecturing. That would be when an instructor spouts lots of words with wild gesturing and says nothing—where explanations and meaning are absent. I certainly didn't want to be an arm-waver in my recitation sessions, or later when it was my turn to teach. Arm-waving at its worst is verbal diarrhea. A particular pleasure shared with Gordon Prather was finding an empty classroom at night where we'd fill the chalkboards with solutions to Jackson's *Classical Electrodynamics* problems. (Fast-forward: It was Professor John D. Jackson who hired me as a guest lecturer at UC Berkeley in 1980.) How satisfying it was to master a subject I had failed in my sophomore year at LTI. Learning can be exhilarating. I looked forward to teaching electromagnetism when I became a teacher. I love explaining electricity and magnetism.

Like most universities, USU attracted celebrity speakers. One was famed TV physics educator Julius Sumner Miller who mesmerized general audiences with physics demonstrations. His lecture was simple in design. He assembled a series of demonstration apparatus along two lecture tables that covered a wide span of physics topics. At one end of the table he showed

Flamboyant Julius Sumner Miller

demonstrations familiar to most of the audience, but each with insightful narrations about the physics involved. Even the simplest of physics can be awesome when articulated clearly. Julius began with inertia demos, progressed to electricity and magnetism, and concluded almost an hour later with light and color. His tactic of raising questions, then with some time for thinking by his audience, skillfully answering his own questions, taught a lot of yum elementary physics. If Julius Sumner Miller wished to influence new teachers, he succeeded big time with me. Due to his inspiration, demonstrations would be central to my teaching.

Later years enjoying the buoyancy of the Great Salt Lake

Utah is an outdoor paradise, especially for those with the time and finances to fully enjoy it. Although a hobby I very much enjoy is skiing, being short on both finances and time I never skied in Utah's great skiing areas. Nor did we tour its beautiful parks. We did, however, spend occasional weekends in beautiful Logan Canyon, and some time went up farther to Bear Lake to join Huey in his love of fishing. We enjoyed some snowshoeing with the Johnsons, and took a swim in the Great Salt Lake, which was full and salty at the time.

One of the bonuses of USU was its proximity to Pocatello, Idaho, home of Millie's brother George Luna and wife Louise. Louise had two children by a previous marriage, Perky and Vicky, and soon after, their own little daughters Georgia and Corine arrived. We spent many weekends with them. This was at the time the Beatles came to America. For us,

The Idaho Luna family: Vicky, Georgia, Louise, George, Corine, and Perky

those were very memorable times. All our Idaho weekends were wonderful.

An activity I enjoyed with the kids was pretending I was a "windy tree." I'd stand in a crouched position, feet properly spaced on the floor, with hands above my knees and challenge kids to climb up my body and try to surmount the windy tree. In a swaying motion I'd say in a moaning tone that nobody could climb to the top of the windy tree. None of the Luna kids could do it. Perky, the oldest, could almost get a leg upon my shoulder, but then I'd gracefully shift position where he'd slide, and gently fall to the floor. I never tired of playing the "windy tree" whenever children were around. Many years later, on June 26 in 2005, my grandson Alexander Hewitt was the first to successfully climb the windy tree. With my skills of bobbing and weaving diminished I leave it to him to continue this family legacy. We're lucky to have our time under the Sun—and know when that time has run its course.

One morning while living in graduate housing at USU Millie and I took an amorous shower together. Again, our efforts at birth control were botched. Nine months later on August 4, 1963, James Kyle Hewitt was born. By this time we were living in brand-new graduate housing with minimum furniture because we'd soon be going to California. Jamie's crib was a cardboard banana box. Like our other two kids, he was a healthy baby with the usual mishaps and doctor visits for usual childhood maladies. Jamie nicely added to the family. Although all three children came earlier than wished for, once they were part of the family, they were much loved. When Millie and I had to leave the kids with brother-in-law George in Pocatello for a couple of days to check out prospects in California, my diary on February 24, 1964, says: "House seems so empty without the kids!! They are such a big part of our lives, which becomes strikingly evident in their absence. Love 'em!!"

Millie with James

Today I like to tell students new to a college or university to seek out the very best professors and take their classes. My best professor at USU was Helmut Hoffmann, chairperson of the psychology department. Not only did I sit in on his lectures, I purchased a tape recorder to record them and keep like one keeps a well-loved book. Professor Hoffmann was a German, who as a very young man was a soldier for Germany in World War II. His slight accent and his voice were what professors would wish for. In a word, he was dynamic. As good luck would have it, America brought Wernher von Braun to lead its space program. I felt USU was similarly fortunate to enlist charismatic Helmut Hoffmann into its faculty. I was very lucky to become his friend.

But I screwed up my friendship with Professor Hoffmann badly in my second summer in Logan. He asked me to care for his home lawn while he and his family were away for a month. A simple request, and I was up to honoring it. My mistake, for which I remain ashamed, was that I put off mowing the lawn until a couple of days before they returned. Oops! Logan was drenched in an unusually heavy rainfall. The lawn mower wouldn't cut the long wet grass. I frantically tried cutting it with a sickle. I failed. When the Hoffmanns returned and saw the tall grass, they, and I, were devastated. He was sensible enough to cancel me from his list of friends. I made no excuses. I violated the trust of a man I felt privileged to know and admire. The message to this misadventure is a well-known one: Don't put off what you should do today until tomorrow.

The summer of 1964 was coming to an end, when I'd be granted two master's degrees—physics and education. Before that I accepted a flight from Utah to California for a job interview at Lawrence Livermore Labs. I did well in the interview and was thereafter was soon offered a position. I turned it down. I wasn't interested in bomb development, but I used the Livermore plane trip to access a prearranged interview for a more interesting position—teaching physics at Cabrillo College in Aptos, many miles south of San Francisco. My interview at Cabrillo College was a disaster. I had rented a car and drove to Aptos on a Saturday for the interview. Arriving at the science building in pouring rain, I knocked at the door and an instructor, I don't recall his name, answered and told me that although the position was advertised, the department had already selected a candidate. I was not even asked inside the building. I was devastated.

I drove north to Mill Valley to spend the weekend with the Johnsons before returning to my family in Utah. Huey suggested that since I wanted to locate close to San Francisco that I should reapply to CCSF. No way, I said, for I didn't wish to encounter Dean Mayo again. Huey didn't reason with me or in any way try to convince me to reapply. He simply picked up the telephone, dialed the CCSF switchboard and asked for the dean of instruction. He was connected to Lloyd Luckmann, *another* dean of instruction. Huey told the dean that he had a friend wanting to teach with a brand-new master's degree in physics, then he passed the phone to me. The next day I was in Dean Luckmann's office and the interview was an opposite experience to the one I had earlier with Dean Mayo. I spoke with Art Austin, the department chair in physics, who told me that a one-year substitute position was open that would lead to a full-time position if my teaching was successful. I ecstatically accepted the position, for it was a dream come true. Huey had once more "rescued me."

In the fall of 1964 I began my teaching career at CCSF.

EIGHT

Teaching at City College of San Francisco

I n late August of 1964, one day after officially receiving my two master's degrees from Utah State University, we packed our belongings in a used trailer and pulled it with our 1954 yellow Plymouth from Utah to California. Our goal was a probationary long-term substitute teaching assignment at City College of San Francisco, which would become a permanent position if I passed muster. Years of educational preparation were joyfully coming to fruition.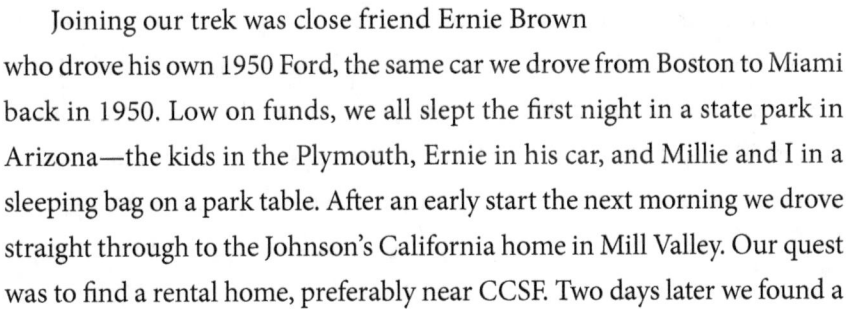

Joining our trek was close friend Ernie Brown who drove his own 1950 Ford, the same car we drove from Boston to Miami back in 1950. Low on funds, we all slept the first night in a state park in Arizona—the kids in the Plymouth, Ernie in his car, and Millie and I in a sleeping bag on a park table. After an early start the next morning we drove straight through to the Johnson's California home in Mill Valley. Our quest was to find a rental home, preferably near CCSF. Two days later we found a house at 1046 Skyline Drive in Daly City, a bedroom community that adjoins San Francisco. With financial help from Ernie, we leased it for a year at $170 per month. My diary entry for that day, August 30: "The good life here at last!!"

CCSF is located in the southern part of the city away from the sightseeing areas and tourist attractions. Set on a hill that is frequently windy, the campus is often overcast and chilly yet quite comfortable weather. In 1964 its modest-sized campus contained the Science Building, the largest classroom building. Two other buildings on the campus were Cloud Hall, which housed the library and additional classrooms, and the administration building, soon to be named Conlon Hall in honor of Dutch Conlon, the president who hired me. A three-day faculty orientation began on September. I had found my personal heaven.

One of my first tasks was to assist in photographing incoming new students as part of their registration process. As each student was photographed I said, "Smile, you're at Silly College." I had the biggest smile! Later the registrar commented that the new group seemed the happiest in college history. The world was mine, for I had my own desk in an office that I shared with another teacher, George Hutchins, ample parking in the faculty parking lot, lights and heating bills paid, all the chalk I needed, a mimeograph machine for student handouts and tests, freedom to teach as I wished—let

Faculty portrait

My office S170 in the Science Building

me say it again. I had found my personal heaven. I remember leaning back in my swivel chair, putting my feet on my desk, and telling myself that all the hardships previously endured paid off. I had made it—big time. No more working with solvent fumes in silkscreen factories, painting signs in bad weather, enduring the uncertainty about earning a living. A wonderful bonus was being paid a full-time salary basis of $7000 a year no matter how many hours or units I worked. I had escaped the working class. I had much more than a job—I had a *profession*, and a worthy one.

The physics department at City College was made up of a stimulating group of teachers. Unlike some other departments, there was no "dead wood" in our department. When I came on board in the fall of 1964 there were Art Austin, Gladys Luhmann (PhD in physics from UC Berkeley), Jim Ripley, George Hutchins, Lance Rogers, Norman Easter as storeroom manager, and newly hired Oleg Rheott. All were outstanding teachers. Oleg had a policy I admired. He gave out lecture notes before each class so that students could follow his lectures without the need to take notes. We addressed one another as Mr., except for Gladys, the only one with a doctorate. So it was Mr. Austin, Mr. Ripley, Dr. Luhmann, and so on. Except for Gladys, we all wore neckties. We made no distinction between being a teacher or an instructor. Either was fine. Years later new staff were PhDs, and modestly went by Mr. also. Lance Rogers was the exception. He earned a doctorate in education at UC Berkeley. From then on he was Dr. Rogers. He relished that.

The first new hire after me was talented teacher Dan St. John, who transferred too soon to administration due to his familiarity with computer programming. Then followed Jerry Hosken, Jim Conley, Annette Rappleyea, and Jim Court, all accomplished instructors. Marshall Johnson came next, but left teaching after three or four years. Ed May followed, and did the same. Dave Wall, Chelcie Liu, and Norm Whitlatch were the last to come aboard. Norm went on to elevate the role of the laboratory in physics instruction for three decades. It was gratifying that our department was staffed by a very competent group of educators. We also welcomed Frank Koehler, a student in my Physics 4A class during my second year at CCSF who continued his education and returned as an evening instructor of the same course! Frank enjoyed what he deemed the perfect situation—doing research at a laser firm in the day, and first-rate teaching three evenings a week. Later, good friend Tsing Bardin became a part-timer, about the time Will Maynez was hired to upgrade the physics shop and storeroom. Mainly due to the human quality of my CCSF colleagues I never seriously considered jumping ship during my stints at guest lecturing.

A happy guy!

CCSF had a dress code for instructors. Men wore jackets with a tie (no jeans), and women wore blouses or sweaters with casual dresses or skirts that fell just below their knees. Our Dean of Women, Mary Golding, was very strict about proper dress. I was aghast one day on the front stairs of the Science Building to see Dean Golding running furiously down the lawn to a bus that was about to unload middle-school students for a tour of the college. The teacher wore slacks, a big no-no!

Dean Golding instructed the teacher to get back on the bus with her students and leave the campus—at once! Oops. This I saw as going too far with regulation enforcement. Except for Dean Golding, I had a positive impression of other administrators and faculty. I was glad that Dean Golding soon retired.

In tune with the Army recruiting slogan, "Be All You Can Be," I wanted to become the best teacher possible. I didn't want to be better than my teaching colleagues, which is a different story. If that were my goal it would be met by resentment. No, I wanted to be the best that *I* could be. Again, very different. For the students that would mean giving my clearest explanations of physics, making fair assessments, and getting graded papers back quickly. I remember looking out my office window at a student asking an instructor when graded exams would be returned. The question had merit, but was met with a harsh response from the irritated instructor. I felt the instructor was out of order. A student shouldn't have to beg for the graded exams. That event would not occur with me! To be all that I could be meant a prompt return of papers. Likewise with grade submissions on time to registration clerks. Being always on time was appreciated. No chip on the shoulder of this guy! All went well for me.

An incident occurred that nearly dashed my teaching career. One evening our new next-door neighbor John Martinez invited Millie and I out to dinner at an upscale restaurant in downtown San Francisco. As we were finishing our dinner, he remarked that we should walk quickly to the restaurant exit as soon as he arrived there. When we did, he quickly whooshed us out the door and into his car parked across the street. Uh-oh, he was exiting without paying the bill. He felt gleeful about this. Millie and I were outraged! Here I was, with all my years of toil to attain my probationary spot at CCSF, all on the line. We could have been arrested and my career would be gone. A lesson learned—be careful who your friends are. We remained distant from John after that incident. Whew!

Along with my teaching profession, I had a family to care and provide for. So on weekends I effectively donned my family cap. On weekdays I wore my teacher cap. Each influenced the other. For brevity, this book concentrates on the teacher side of my life, which was more than fine in spite of being dependent on bank loans and enduring continual financial difficulties. On December 26, 1964, after a joyous first Christmas in California, my diary states: "I am a very wealthy person. I have my wonderful little family—my

Paul   Leslie   James

Hon, my Paulie, my Leslie, and my little Jamie. As long as we are together, I shall always consider this a wealthy family." Wealth was more than finance. Despite financial difficulties we enjoyed the best recreational opportunities that the Bay Area offered. Family and great friends—what could be better. With Huey and Sue Johnson we took in lakes, parks, and redwoods on the other side of the Golden Gate Bridge. We even went to Disneyland in Los Angeles during a semester break.

Leslie was as model a girl as parents could hope for. Paul and James, on the other hand, experienced sibling rivalry, with older Paul too often bullying younger James. I was proud that little Paul created a neighborhood newspaper that he distributed to friends. All three kids went through the stages of raising hamsters and other pets and succeeded just fine with neighborhood interactions. They progressed with little or no serious childhood problems.

Being content with the size of our family I later got a vasectomy, a simple medical procedure to assure total birth control. The procedure could be reversed if we had a change of mind. That didn't happen. We were a happy family.

Our rental home in Daly City was fine, but we wished to purchase a permanent home. On May 23, 1965, our search for that ended when we found 667 MacArthur Drive for sale in the Broadmoor district of Colma, just a five-minute drive from CCSF. It was a three-bedroom split level home in excellent condition. A big plus was the upscale Garden Village Elementary School for the kids within walking distance of our home. Most importantly, it had a huge back yard, just perfect for our family. It even had an adjoining "in-law room," amenable to visits by Millie's family from Colorado. The home was priced at $24,500, and a GI loan made the purchase possible with no down payment. In fact, the monthly payments for the home were $171, one dollar more

Finally our home in Broadmoor Village

than the rental for our Skyline Drive home. We directed our attention into improving our new home by painting both interior and exterior surfaces, building a stone fireplace, installing a heater in the back room, and even building furniture for the kids' rooms. Again, life was good!

A few years later we added to our home when carpenter Kirby Perschbacher came from Colorado to visit James. They constructed a lovely ten-foot-tall pyramid on the roof of the back bedroom, complete with a large triangular window looking out at the eucalyptus trees in the back

Pyramid atop the back room

yard. James was delighted with this addition to his bedroom.

My earlier thinking in graduate school was that once I began teaching with a steady income, our financial problems would be minimal. This turned out to be very wrong. Home expenses, kids getting sick, house and car repairs, and unexpected bills posed the same financial difficulties experienced during student days. To make ends meet between semesters I was lucky to get part-time work cutting silkscreen stencils at Ray Advertising in downtown San Francisco. Additional income also came when I accepted a $780 National Science Foundation grant to attend a six-week summer session for teachers at the University of California, Berkeley.

At Ray Advertising I met a new friend, Paul Olson, a stimulating and talented artist. Millie and I shared many good times with Paul and his wife Emily. They introduced me to a Bob Dylan concert at Masonic Auditorium in San Francisco. Wow! I'd never seen an audience of so many weird looking people. A sort of social insurrection was happening—and it was stimulating. A week later I attended another wild concert at the Fillmore Auditorium, a benefit for the San Francisco Mime Troupe hosted by Bill Graham (not to be confused with the evangelist Billy Graham). My world was widening.

Then in mid-December Paul Olson phoned me saying he had a substance called LSD that expanded thinking. I had just read about this in *Time* magazine—how a professor Timothy Leary at Harvard touted the wonders of this new chemical. Expansion of thinking? I saw investigating LSD as my responsibility as an educator. Paul invited me to his home and we each tried half a pill. WOW! As I wrote in my diary, "pleasant, peace, bewilderment, fantasy!" This was an almost eight-hour extraordinary experience. How I wish that Millie could have gotten a babysitter to join us. That came later. I asked if there were something less potent with less time involved. The answer was yes—marijuana. Uh-oh, that's what I politely refused the previous year in Pocatello when my brother-in-law George welcomed me into the family. No dope for me!

The times were changing not only locally, but gradually across the whole country. *Psychedelic* became a new word in conversations and a buzzword on television. Millie and I began experimenting with marijuana, and occasionally with LSD. The phrase "turn on" seemed appropriate with LSD. Neural activity that dealt with time and intensity of experience were turned up full blast—enormously more than used normally in everyday activity. The mind experiences *information overload*—big time. This could be quite troubling to those with questionable mental health. LSD is not for everyone.

An important feature of LSD is surrendering of control. You can't decide what you want to trip on. When I wished for new insights into special relativity, I read a lot of material about it before taking LSD. It didn't work. I couldn't bring relativity into my thinking as my mind went elsewhere—completely beyond my control. Millie and I didn't overdo our psychedelic experiences, which to us were recreational encounters, and with LSD, always with a "sober" friend present. We never experienced a "bad trip," as was often reported in the media. Heroin and addictive narcotics were a big no-no. Like our friends, we were recreational moderate and usage tapered to zero over time. Legally, smoking pot went from toleration to being a crime. Many went to prison for what we did. We kept all this to ourselves and our close friends.

In 1966 my teaching salary was raised to $8030, and an extra $9.45 per hour for teaching an overload in evening classes. Nevertheless, financial stress was still a problem. Loans at Beneficial Finance were frequent. Credit cards were not yet in vogue. I also managed to take a train trip to visit my mom and brother Steve in Texas who had some personal problems that fortunately worked out. I tried my hand at painting pictures with frames I made in the physics storeroom shop and gifted them to my mom, sister, and brothers. And then there were summer road trips to Salida, Colorado, enjoying friends and family while being busy painting signs. All this was the background to the adventures experienced at the onset of my teaching career.

At a CCSF faculty party welcoming new recruits that involved a fair amount of drinking, I was bothered by an instructor who boisterously claimed that we should all be honest and admit that we're teaching for the *money*. After glancing at my expression of disbelief, he muttered something to the effect that he was speaking for *most* of the teachers anyway. He certainly wasn't speaking for me, or any of my colleagues in the physics department. We all had a passion for teaching that wasn't connected to our salaries. Our reward was the sheer satisfaction provided when motivating student learning—lifting our students educationally. Achieving our best at teaching was rewarding enough.

I greatly respected being on the faculty of CCSF. Its great tribute is that students who graduate and transfer as juniors to difficult-to-be-accepted-UC Berkeley outperform the UC Berkeley students as juniors. Although brighter students on average enter UC Berkeley, the higher quality undergraduate teaching at CCSF finds students better prepared than their UC Berkeley counterparts in the junior year. This well-deserved academic reputation of CCSF kept me there for thirty-two years, along with delightful leaves of absence to guest teach at major universities in California and Hawaii.

In addition to algebra and calculus based courses, I was assigned to teach the course for nonscience students, Physics 10, *Descriptive College Physics*, named after the popular textbook we used that was authored by Harvey White at UC Berkeley. Only department head Art Austin and I taught this course. Other instructors preferred to teach the courses for prospective scientists and engineers. Teaching nonscience students was nicely in line with my mission: to lift student expectations of themselves as physics students and gain a wider respect for science, enjoyably. I wished to guide the joy that occurs when students discover they understand what they thought they couldn't. For example, if you're at a lakeside tossing rocks into the water and someone tells you that somebody got eight skips with a flat rock, and you accomplish nine skips—you're happy. Achieving more than you expected makes you happy with yourself. Being the teacher to elicit that joy made me a happy guy.

Substance is also important. I framed a study of physics as a study of nature's rules. I impressed on students the value of knowing these rules by comparing the importance of knowing the rules to sports or even a party game—that not knowing its rules means not appreciating the game. And so it is with physics—we can't fully appreciate our everyday world unless we're knowledgeable of nature's rules. But most of all, my teaching quest was to show that science, and physics in particular, can be a lens through which students not only appreciate the world, *but make sense of it*—in contrast to the all-too-popular anti-science explanations. My message was relevant at a time when a growing counterculture was condemning science and technology for the ills of the world.

Teaching lab

To beef up my lectures I shared some of the best 16-mm Physical Science Study Committee (PSSC) black and white films with my classes. These were short lectures by noted physicists such as MIT's J. R. Zacharias, Stanford University's Albert Baez, and the University of Toronto's entertaining Patterson Hume and Donald Ivey. All films featured physics demonstrations presented by experts. I showed perhaps a half dozen at the outset of my teaching, then less of them as I mimicked the experts in their ways of showing demonstrations. Ones that couldn't be duplicated in class were the feature of the last PSSC film I shared with students—the classic *Frames of Reference* with Hume and Ivey. Upside-down camera shots weren't something a teacher could mimic in class. Another was the insightful twelve-minute animated *Mathematical Peep Shows*, by Charles and Ray Eames. These films were a springboard to improving my presentations. Like training wheels on a child's bicycle, they could be removed as teaching skills increased. Fast-forward decades later and my *Conceptual Physics Alive!* video lessons nicely serve the same function in many classrooms worldwide. I'm very pleased with that.

I soon found that
making physics appealing
to my nonscience students
was in conflict with the
computational examples in
the Harvey White textbook.

Most of my students had an aversion to *numbers*. So I focused on teaching concepts qualitatively. Proportions and equations were most often first written in English. After some familiarity, I advanced to shorthand symbols. For example, to illustrate relative magnitudes of velocities involved in momentum changes for a cannonball shot from a cannon, I'd draw the symbols for masses and velocities in different sizes to indicate their magnitudes. They were learning the math of physics conceptually.

Although I devoted zero time to computations, my focus was on equations without numbers. For example, I'd begin a lecture on gravity by citing how wonderful it is that music can be expressed as a pattern on paper—a musical score. Likewise, the wonder of the universe can also be expressed as a pattern. How might the pattern appear that underlies the stars shining in the nighttime sky and all the cosmos? Patterns underlie both music and physics. I teased that for fifty cents each, I'd reveal the pattern that underlies the universe. With interest piqued, I wrote Newton's law of gravity on the board, as a proportion with $F \sim$ rather than an = sign.

Equations can express patterns. They show the links between concepts. Thus, equations were a big part of my teaching, not as recipes for solving physics problems, but as guides to thinking. Reading equations is akin to reading notes in a music class. Their value extends beyond the classroom. We're all familiar with many people who are convinced of an idea by not knowing more than one side of the story. Aha, physics equations have more than one symbol and incorporate multiple factors to arrive at a solution. When this skill of thinking of more than one thing carries into everyday life, hooray! Only later do I introduce G, the constant that converts a proportion to an equation. Equations not only guide good thinking, they promote broad-mind thinking.

Equations complement "What if" questions: What happens to the gravitational force between two planets if the mass of one of them doubles? If both double? After many such questions and with some fanfare, I'd place a period at the end of the equation. That tells us that only masses and distances play a role in gravity. No symbol for speed, color, or temperature meant they were not part of the picture. Equations are wonderful guides to thinking about what is, and what isn't, part of physics. Want to plug in some numbers? Not in this course, for time is better spent moving onward.

My teaching was not without difficulties. In 1964 entry in my October diary states: "I'm not the superior teacher I thought I would be—not really doing a good job—will try harder." And then on October 20, "Teaching is hard work after all—making up tests, correcting papers, reports and preparation, consume much time—I make goofs in class too—polish is quite a way off, I'm afraid." Both a dean and Art Austin observed my classes, and didn't share in my pessimism. Perhaps I was being too hard on myself. On November 4, "Class went well—have renewed confidence in my teaching."

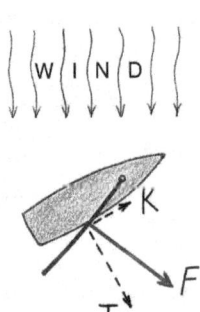

But later when I attempted in an evening class to explain how sailboats are able to sail into the wind, I made a terrible blunder. I mixed force and velocity vectors on the chalkboard—a huge no-no. One never combines a velocity vector with a force vector!

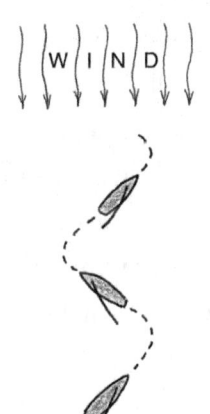

My carelessness emerged in that lecture, which I'll never forget. Embarrassingly, I promised I'd correct my explanation in the next class meeting, which I did. So my teaching performance was sporadic. Teaching has an enormous advantage that other professions don't: No matter how you screw up, students in the following semesters don't know about it. You start each semester with a clean slate. My teaching benefited by this, for I think I must have made about all the classroom mistakes that new teachers can make. But the mistakes are not cumulative. As said, we begin anew each successive semester.

Deciding what material to exclude from a course is as important as deciding what to include. A new teacher fresh out of graduate school faces the task of condensing years of study into a sixteen-week experience. What to *not* cover? For a nonscience course, certainly don't spend time on the tools of physics that are essential for physicists, but not for nonscience students. Don't spend time on significant figures, graph construction, converting coordinate systems, and distinguishing units of measurement, and most certainly on the oodles of techniques for solving problems. Including these in a nonscience class would be worse than a waste of student time. For many, it would be a dreadful experience. I vowed for years NOT to inflict that on my students.

Also important is the academic plow setting—how deep the physics goes. Another is learning to distinguish concepts—what is useful to learn and what is not. I was not in the "no pain, no gain" club. My club was "let's make a student's first course in physics a delightful experience." Different strokes for different folks!

I also heeded the dreaded "information overload" that I endured at Lowell Tech years before. I focused my lectures on only a few topics, and sometimes a single one, with real-life examples. I remember Art Austin in my hiring interview asking how much emphasis should be given to elementary particles, one of the talked-about issues of the day. I replied that before my students were taught about the paths of particles in cloud and bubble chambers, I'd want them to know what produces the bubbles in boiling water. The physical process of boiling water for a cup of tea was a more important topic than the particle tracks in the chambers. Art and I were on the same wavelength, emphasizing the physics that surrounds us in the everyday world.

Art was very accomplished in his teaching and was very well liked by his students.

Teachers who focus only on content find that teaching the same material over and over becomes tiring. I think of my friend and colleague Dave Wall, physics teacher and magician, who was asked if the trick of pulling coins out of little kids' ears was getting old for him? Dave replied not at all, because he had a new kid each time he did the trick. Likewise, teaching Newton's laws over and over isn't tiresome because we have a new class each and every semester. Although concepts being taught were learned long ago, a good teacher relishes in sharing that knowledge—semester after semester. Too many teachers who lack this attitude for teaching soon burn out and quit the profession.

To improve my lecture presentations I tape-recorded them with a small portable recorder at one end of my lecture table, which provided me feedback to my classroom presentation. Although listening to my lectures at home was time consuming, I benefited by correcting what I heard as quirks. I found that I said "okay" too often. What seemed okay to me was often frequently not okay with most of the class. A more important discovery was beginning

Always a lecture hall of attentive students

a topic, taking a question from the class, and then not returning to the topic after giving an answer.

An incident in one of my classes is worth mentioning. I posed a particular question to the class and a student gave a puzzling but insightful answer. When I complimented him he asked why I didn't explain it as he did. I responded that his explanation was better than mine because he was likely smarter than me. The class was shocked. Then I explained that several students in the class maybe more intelligent than me. I'm supposed to be the most knowledgeable person in the room, not necessarily the brightest. It's true. We should be okay with that.

I taught a variety of courses in addition to Physics 10. Whatever my schedule, I taught my favorite Physics 10 on Wednesday evenings. The time gap between afternoon and evening classes resulted in something wonderful—*my daily nap*. Even today, I sleep in bed at 4 p.m. for an hour, convinced that this daily siesta-like routine contributes significantly to my good health. An added bonus is often waking from the nap refreshed with good ideas. Cheers to a daily nap!

The performance expected in conventional physics courses did not apply to my Physics 10. My goal was to instill a motivation for learning. For all the courses I taught, preparation was greater than I expected. I vowed not to be a teacher who would "wing it" in class. Physics 10 being a one-semester course meant sixteen weeks to promote my passion for physics. I decided to lessen the quantitative tone of White's textbook by supplementing it with a paperback lower-level book, *Vitalized Physics*, by Alexander Taffel. Teaching continued to be enjoyable.

I attended the National Science Foundation six-week summer session for teachers at UC Berkeley. It was less than satisfying. About thirty teachers attended daily lectures about thermodynamics, much like a course in grad school. Nothing from that experience contributed to any improvement in my teaching, and I think the other attendees would agree. The focus was a trial for new books that the Berkeley physics department were producing for advanced undergraduate courses. How much better it would have been, I thought, to instead use the same textbooks that instructors might actually use in their own teaching, and show effective ways of presenting material. Perhaps this would seem too "low level," since the purpose of the session was to expand knowledge, not to more effectively teach what attendees already knew. The best part of the experience was becoming acquainted with other new teachers as well as some professors in the UC Berkeley physics department.

I was usually the last teacher in the physics department to leave in the afternoon, mainly because I graded lab reports alone in the classroom rather than taking them home. One afternoon Art Austin watched me poring over student reports, writing comments in red ink on each report. He was not impressed with this personal attention and advised me instead to "spot check" the reports. He said that my energies would be better directed to lecture preparation than to excessive time in grading lab reports. I took his advice. I'm convinced that no students suffered by the spot-checking method. The stress of teaching was lowered, which was good for all concerned. I was loving everything about City College!

I should say *almost* loving everything. I participated in a campus-wide faculty hiring committee and observed faculty interviewing teacher candidates. The faculty would select two candidates to see the president, who would make the decision. I noted something disturbing in the faculty interviews. My idea of who would be a good teacher was very different from those of the interviewing faculty. Dynamic candidates were passed by for weaker but straitlaced "safe" candidates. Could it be that teachers didn't relish new colleagues who might outshine them? Jealousy or envy can sometimes play a role. Unfortunately, my view of this intensified over time.

The hiring of first-rate new faculty is perhaps the most important administrative task of a department. Long before the hiring of brilliant PhD instructors Chelcie Liu and Tsing Bardin, there were no Asian teachers in our physics department, although there were many Asian students in our classes. To correct this, we wished to hire an Asian instructor when a position became available. This occurred when a Chinese woman applied while

I was abroad. We had a phone interview and I was bothered that she didn't seem to know her physics. When I asked why Earth gravity doesn't cause satellites to crash into Earth, she replied that the gravitational pulls of stars and other satellites prevent this. Whoops, she made similar goofs with other questions. My vote for hiring her was a no. Others felt otherwise and she was hired anyway. Being both a woman and a minority was too great to pass up. As it happened, she taught for one year and then accepted another position in Southern California where she wished to live.

To convince my colleagues that hiring was especially crucial when hiring minority, I made this claim: the choice of a new faculty was between a mean-spirited brilliant Nazi with a swastika tattooed on his forehead and a weak bozo minority candidate, CCSF would best be served by hiring the Nazi. Why? Keep in mind the thousands of students influenced by these two applicants. The Nazi candidate might degrade Nazis, which I think we can agree is preferable to presenting a weak minority teacher to students. Hiring a minority teacher with classroom shortcomings would not only be counterproductive in lifting student aspirations, but would tend to confirm a negative stereotype to thousands of students in general. All students, minority or otherwise, are best served by competent teachers, and all the better if they happen to be a minority. To me, the argument to hire only an excellent minority teacher has merit. To others, "Hewitt wants to hire Nazi teachers." Give me a break! As it happened, due to the fervor of hiring minorities, the next two Chinese applicants, quite less than brilliant, were hired—and as time would tell, the last to retire. I'll leave it at that.

My growing popularity at City College became problematic. It was difficult for Art Austin when students requested a transfer from his section of Physics 10 to mine. Transferring was common and not unusual for a few students. But transfers to Mr. Hewitt's section expanded. The explanation may have been as simple as this: Mr. Austin was an old man talking. Mr. Hewitt was a younger man talking. Students wanted to be in the younger man's class.

Art prided himself with his excellent teaching. Why were students transferring? Thinking that I was lax on grades, he inspected the grading criteria of a recently hired astronomy instructor who filled classes by easy A and B grades. To bring administration's awareness to my grading (which always averaged C), Art had to confront the new astronomy teacher. Uh-oh! The new teacher fought back and attacked both Art and the college administration full force. I remember Art taking medications for stress, something he hadn't done in the past. Finally in about a year or so, Art took an early retirement, a broken man. This was a big setback for the otherwise normally functioning physics department.

Art Austin made no secret of his distaste for bearded men, who he saw as shiftless and untrustworthy. After Art retired I and two other instructors let our hair grow and wore beards, which were popular at the time. My thin chin, skinny neck, and protruding Adam's apple were nicely covered. Although I still weighed about 115 pounds, I felt better with my longer hair and beard. Ichabod

Classroom demonstrations were common

Crane transformed into a more respectable looking Mr. Hewitt. It was wonderful to be among the normal looking people.

Following Art's retirement, I taught all sections of Physics 10. Lecture halls were filled to capacity, for counselors continued to recommend Physics 10 to incoming students, viewing it as a great bridge between high school and college. With five full sections of Physics 10, two of them held in the 320-seat auditorium in the newly built Visual Arts Building, more than 1000 students enrolled in Physics 10 each semester. There were fourteen instructors in the physics department, including part-time instructors, serving a total of about 2000 students per semester. Mr. Hewitt was serving half this load. No elected course at CCSF drew a consistent 1000 plus students per semester. I was appreciated by both the administration and the students, receiving from the CCSF Student Council a plaque for being an outstanding teacher in Fall 1967.

On April 4 in 1968 Martin Luther King Jr. was murdered. Out my office window I saw how large the crowd was lamenting this national tragedy. This was a troubling time nationally. Following this on June 5 of the same year presidential hopeful Senator Robert F. Kennedy was shot as he walked through the kitchen hallway of the Ambassador Hotel in Los Angeles. Again, a large rally assembled in the CCSF courtyard. Sometimes I paid too little attention to events on the outside when my efforts were indoors at CCSF.

In May 1968 the headlines of *The Guardsman*, the campus newspaper, was "Hewitt Lauded As Outstanding In Evaluation." This was the first year that the Associated Student government organized a campus-wide assessment of instructors. This was also the time that students across the country were rebelling

Displeasure with the Vietnam War

against campus practices and the Vietnam war. The Vietnam War was in full blast and protests against it were everywhere. I was among the protestors for peace and marches were commonplace with me. Some of my physics students and I joined the 100,000 protestors in the huge Moratorium Peace March in November of 1969, displaying a banner "Physics Students for Peace." Although some anti-war activity occurred at CCSF, most centered on the nearby San Francisco State University campus.

College politics were undergoing changes in many parts of the country. At CCSF students began to demand fairness and a voice in campus affairs. This resonated with me. When I first came to CCSF and participated in the faculty hiring committee I suggested student feedback as part of instructor assessment. This was vehemently protested, especially by some unpopular instructors who felt students only appreciated softness. Besides, it was argued, students don't have the maturity to judge the value of their instruction, and cases were cited where instructors were appreciated only years after their courses. In short, student input to evaluation didn't happen—until the fall of 1968.

One of the instructors in the biology department that students loved was Darwin. Nobody was more friendly to students than Darwin. But he was crushed to find a low rating in a later student survey. Comments were that although students liked Darwin's friendly and fair manner, they didn't learn from him. Afterwards, Darwin wasn't the upbeat fellow in the hallways. I've since thought that how much better it would have been for Darwin to discover his weaknesses early in his career, when he could easily have made corrections. But here he was, nearing retirement, finding himself unappreciated. Had student assessment been in vogue when I came to CCSF, I'm confident that Darwin would have retired a happy and contented man.

I am thankful that CCSF was ahead of the curve with respect to addressing and lowering social problems. Freedom of expression was not suppressed—not quite, there was one exception. Our dean of students, Ralph Hillsman, was adamant that a free speech platform not be constructed on campus. This disturbed me, because Ralph had always promoted fairness in student affairs. But times were changing, and against his wishes, a free-speech platform was erected near his office at Conlon Hall. It faced the huge empty lawn extending up the hill to the Science Building. I was one of the first to express my anti-war views to hundreds of students from the platform. These views were expressed out of class, and very little if at all in my classes.

The Free Speech platform had multiple uses. I remember seeing the vocalist of a new group "Big Brother and the Holding Company" bellowing her heart out from that platform. It was Janis Joplin, before she became famous. Another use of the platform disappointed me. That's when Jesse Jackson came to campus to address hundreds of students, mostly black. I was hopeful his message would be uplifting, proclaiming this time of opportunity for black students to continue their education, make something of themselves, and establish a role in the formation of a better country. But no. Jesse said none of that. Instead of leading followers to higher ground, he in effect placed chips on the shoulders of his audience by reminding them of their slave heritage, stoking the fires of victimhood. Jesse favored entitlement over achievement. My opinion of him as a leader took a complete flip. No longer was I a fan of Jesse Jackson.

Going back to the Vietnam era, some Americans feared a growing communism, not unlike the earlier communist scares that stoked the Korean War that I was a part of. What would happen, they fearfully cried, if Vietnam became communist? The answer came ten years later: Vietnam did become communist and the world did not come to an end. Rather than succumb to victimhood, the Vietnamese rebuilt a prosperous communist society that manufactured and supplied such items as shoes worldwide. Why the war? As comedian George Carlin quipped, that to keep our economy going, every twenty years we have to find brown people to bomb. Supporting this view was President Dwight D. Eisenhower's warning when he left office—the unchecked growth of our industrial military complex. The shame of that war is to be borne by policy makers aligned with the war industry, not the soldiers who honorably put their trust in them. In response to those policy makers it is appropriate that the Vietnam Memorial in our nation's capital is not a high, glorified monument, but is instead low to the ground.

Because the Vietnam War was so horrible, I and others found consolation that it would mark the end of horrendous military misadventures designed to change the politics of sovereign countries. That comfort was brief. Who would know the same military-strength legislators would repeat the same in Iraq and Afghanistan. Is the male urge to fight that's embedded in our DNA something we can't outgrow?

In the next chapter I'll tell more about why Physics 10 became increasingly popular. Much has to do with the first day of class—the most important day in a course.

NINE

First Day of Class

First impressions are often lasting ones. The first day of my course is the most important day of the semester. In addition to stating my expectations of my students, I can set the tone in that opening class—the *flavor* of the course. Early in my teaching when I got my feet on the ground and knew what to teach and what not to teach, I set out to inspire students far beyond their expectations—to inspire awe—to "blow their minds." Here is how I did it.

First of all, my course, Physics 10, was an elective course that wasn't a prerequisite for any later course. This allowed the leniency I desired. I began my class by writing my first and last name on the chalkboard, followed by my personal telephone number. I took the risk of being besieged with phone calls. But that rarely happened. The idea of me allowing phone calls impressed them. I passed out three by five index cards for student names and addresses, a common practice. In addition to the usual names and addresses, I asked them to write at the top right-hand corner of the card the number 1, 2, or 3. I paused, ready for them to ask what I meant by that. I then explained that writing a 1 would tell me that they were fine if numerical problem-solving were part of the course. Writing a 2 told me that they'd be okay with a bit of mathematical problem-solving, but not to go overboard. Those who wrote a 3 indicated they wanted *no numbers* in the course—except the page numbers of their textbook. No numbers! So which is it: 1, 2, or 3?

Upon collecting the cards, I took a minute or so to scan through them. Students were eager for my answer. Then I announced that the course would focus on three! There would be no numerical problems in the course. Any mathematical language would be spoken in English! I explained that there were plenty of other physics courses that used lots of numbers, but not this one. A loud cheer was the response. (No numbers was a bit of an exaggeration, for I expressed the acceleration due to gravity as 32 ft/s^2. Later in the International System of Units (SI), it was 9.8 m/s^2, which I rounded off to 10 m/s^2. Sometimes numbers were helpful.)

To further stoke their awe, I passed out copies of the final exam. That's right, copies of the exam they would take at the end of the course. Looking out the classroom door to see that no authority figures were spying on the class I announced, "I'm on *your* side." I gave them Scout's Honor that I was sincere. I explained that "Scout's Honor" did not necessarily mean I was telling the truth, for the truth that one holds may later be seen as false. Truth is relative. Scout's Honor, with my three-fingered scout sign meant that I sincerely *believed* that I was telling the truth.

The final exam consisted of fifteen qualitative questions. I responded to their incredulous expressions by saying I didn't expect them to answer all fifteen during the final. I'd select four questions. Which four? That would be decided on final exam day! The answers were to be handwritten in examination blue books, which were popular at the time. The final exam score, however, constituted only 25 percent of the final grade.

Final exam questions were straightforward. For example, "Why does warm air rise?" "What role does the greenhouse effect play in global warming?" and "Why is the sky blue?" Questions would cover my minimum expectations of what they should learn in the course. If their physics explanations were clear and correct, I'd be a happy guy. I told my students that my intention wasn't to challenge them by building the problem-solving skills of a conventional physics course. Instead I wished to awaken them to the physics in their surroundings, which did not require laborious homework. Students had enough homework from other courses, but there would be no homework for this course. The notion of no homework elic-ited a cheer. I told them I was gambling that they'd learn a respectable amount of physics

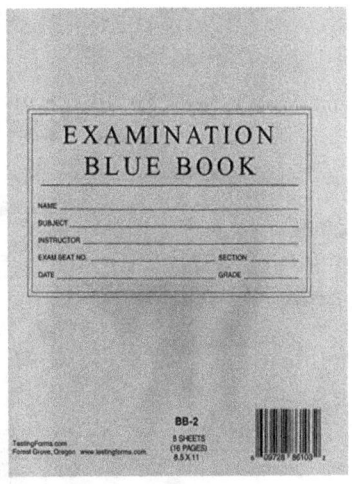

by reading the assigned chapters in the textbook and involving themselves in my lectures. Comprehending content was the name of my game. I wanted them to view the world about them through the physics lens that guided my outlook—a lifelong view, not just a means to a grade as in the many other courses they were taking. No "drill and kill" in this course! I wanted to enhance their way of thinking.

Instead of teaching a bit of physics, I spent much of that first hour asking a series of intriguing physics questions for the class to think about, the answers of which would define what the course is about. For example, why does a hockey puck not need a force to keep it sliding along the ice? We'll see why in Chapter Two. Why does a dropped ball take the same time to reach the ground as a horizontally thrown ball? We'll get to this in Chapter Three. Why does warm air rise? Why doesn't it go sideways instead of straight up? We'll return to this in Chapter 15. Why this, why that, all interesting ques-tions with a promise that all in the class will be quite comfortable with them by the end of the course. I let them know that my ambition was to help them learn to view the world as I had learned to see it. Physics is everywhere. My intent was that they'd come to agree.

Students were joyful when I announced there'd be no homework in the course except for reading the textbook. I told them I was aware that some of

The mix of joy when learning physics

them were already overburdened with other courses that demanded much of their time. It might seem to others that my policy of no homework suggested entertainment rather than learning. I take issue with that. *The physics to be learned is in the textbook.* My job was to *motivate* serious reading of the textbook, and often re-reading. I achieved that motivation by powerfully interesting presentations. "Don't blow it," I'd say, "Take this course seriously! Now may be the only time in your life to learn some physics—which can be a part of your education that you're proud of."

My own pride as a teacher had much to do with my background of hard work in factories and printing shops. From my working class point of view I saw being a teacher an enormously high prize. Would I have valued my teaching career as much if I went immediately to college after high school, and began teaching ten years earlier? I don't think so.

Due to my positive attitude and focus on fairness, my classes were always filled to capacity. Interestingly, my average student grade for Physics 10 over my career was C. Although not a great grade, most students were satisfied to earn a respectable C in a physics course they would otherwise fear. Although the popularity of some courses relates to an easy A, my course was not one of them.

I must say here that the calculus-based courses I taught for prospective physicists and engineers were much different courses. They entailed hard work with very different goals. They were poles apart from Physics 10. But I knew by experience that people can learn a lot by simple exposure to interesting explanations of everyday wonders—the appeal of my mentor, inspirational Jacque Fresco. He opened my mind to *learning*. He assigned no homework—no serious study. I wanted to be a Fresco, and more. Hence the lightness of my Physics 10 course.

An unanticipated bonus of blue-book exams was acquiring a wealth of "wrong answers" for my increasing supply of questions for multiple-choice exams. If there were any banks of good exam questions at that time, I wasn't aware of them. I wrote my own. An exam question having choices that are obviously incorrect is a poor question. It may measure common sense, but not physics understanding. A good multiple-choice question has choices that all seem credible. Choosing between them requires an understanding of physics. The exam should measure learned knowledge, not smarts. Writing multiple-choice exams became an art that was greatly helped by my supply of blue-book student misconceptions. Having a large supply of multiple-choice questions turned out to be essential when class sizes grew and grew. I had to change from blue-book exams to multiple-choice ones with the increased enrollments in following years.

I'd continue that the first day to explain the exams they could expect. There would be three exams during the semester, each with thirty multiple-choice questions. Each exam would count as 25 percent of the final grade. I promised that sample exams would be passed out to the class well before the exam day. In terms of a letter grade, a score of twenty-five or better would rate an A; twenty to twenty-four a B; fourteen to nineteen, a C; below fourteen, a D. An F grade was for those who gave up and didn't complete the course. No student who participated and completed my Physics 10 courses ever got an F. My pitch didn't stop there. I told them that perchance they took an exam on an *off day*, scoring less than they had expected, they could take a similar "retake exam." The scores of the first try and the retake would be averaged. If the average score wasn't to their satisfaction, a student could retake the exam again and again until satisfied. It was clear to all that a B was quite achievable to students of average ability. A few got A, and most seemed content with the traditional C. I like to think that they all had a heightened respect for science in their lives.

All my exams were taken with open books. Sometimes I also allowed open notes. My questions weren't a measure of information recall, but of meaning. For example, how gravitational force changes between a person and a planet when the distance is tripled, is not given in the book. The relationship is stated in the book, but the answer must be figured out. So answering questions went beyond searching for them in the book. Students may have other courses that test what they remember, but in my course, testing went a step further and prodded them to think. It was common for students at the end of the course to say, "This course taught me to think!" We all appreciate using the best of our brains.

Now, how to keep the number of students manageable for retake exams? I did this by citing an example of how averaging might not be a good thing. On the board I'd show that if a student gets a 20 out of 30 questions correct, a B, and studies all weekend and retakes the exam and gets an 18, the average is then 19, a C. Oops, grades can go down just as they can go up! This was enough for about three-quarters of the class to not opt for retakes. The quarter that did opt for retakes made the system manageable. But more important, it gave an unmistakable notice that my course is FAIR. Fairness in grading, more than anything else, is appreciated. Students are entitled to that fairness. I was fortunate to have ample supply of teaching assistants to make this happen. Without them, all would be overwhelming. I think that any teacher using my procedure for first day of class can be assured of great student interest.

There was the perennial problem of students who cheated on exams. To offset this, I instituted the "cheater's corner," on one side at the top of the lecture hall. I explained that I didn't approve of cheating, and my grading policy didn't merit it. I also explained that just as there are people who can't help abusing alcohol and other substances, there were some students who couldn't resist cheating. They were unfortunately "wired that way." My main objection was that honest students shouldn't be distracted at exam time by neighbors attempting to cheat. Herding cheaters to the top corner was effective. Most of the room was quiet, which was appreciated by most students. In time I abandoned this practice, not wanting to seem to condone cheating.

Only quickly drawn art was appreciated

An additional feature was my skill that was nurtured in my silk-screen printing days. That's drawing cartoons while speaking. Interestingly, the number of cartoons that I casually draw is less than a dozen or so for the entire course. Taking the time to perfect this art is well worth it. The chalkboard art to be learned will last one's professional lifetime. Go for it! (One of my favorite activities in later physics conferences is my "Cartoon workshop," where I show teachers standing at the chalkboard how to copy some of my favorite cartoons.)

My course content was influenced by an author's alarming statement in the preface of a physics textbook for a premed algebra-based course: "There is no place for *gee-whiz* physics in a serious physics course." I had the opposite view: I wanted each of my lectures to be highlighted with "gee-whiz" gems and to ideally build up to a "gee-whiz" conclusion. As examples, "A bullet shot horizontally curves and falls to the ground in the *same time* as one dropped straight down from the same height!" "Nobody can remain in the air for more than one second in a standing jump." "A sailboat can sail faster angling into the wind than when sailing in the direction of the wind!" "The atoms that make up your body were once in the bodies of all who lived before you and will live in the bodies of all who will follow!" "Your body emits the same amount of thermal energy as a lit 100-watt bulb!" "Electricity and magnetism connect to become light!" "Atomic spectra are fingerprints that tell us what stars are made of!" "You can never reach the end of a rainbow!" "Earth's interior is kept warm by radio-activity!" "Although nuclear fission and fusion are opposite processes, both can release energy—but not always!" "A traveling astronaut in space ages less than a stay-at-home twin on Earth!" "Shake your book to and fro and you produce a gravitational wave!" Mumbling these statements takes away from their impact. Pronouncing them with gusto can be a great "gee-whiz" addition to a presentation. Delivery is important.

Saved by the inertia of the anvil

Then there are classroom antics I discovered to nicely flavor my course. To illustrate the concept of inertia, I'd lie on the lecture table and ask for a student volunteer to strike a sledgehammer upon a sixty-pound anvil resting on my chest. I was too trusting of a student's ability with a sledgehammer. In 1967 the student volunteer missed the middle of the anvil and smashed the fingers of my left hand. Uh-oh. Off to Kaiser Hospital where fingers were repaired and put in a splint. This was astonishing to the class—and to me! A few days later I had an excuse to see CCSF President Harry Buttimer. "Look at this Harry, a student in my class broke my fingers with a sledgehammer. These kids are getting out of hand these days." That was good for a laugh. In future anvil-sledgehammer demonstrations I had a storeroom employee with sledgehammer skills pose as a student in my class. When I called for a volunteer, I selected an imposter for safety.

Another antic was illustrating center of gravity. An object standing at rest will remain in equilibrium if the center of gravity lies above and within the support base. For a standing person the support base is defined by the perimeter of their shoes. This means that a person standing against a wall cannot bend over to touch their toes without toppling—unless the support base is unusually long. I demonstrated this by attempting the same in class. It can't be done, I claimed. In fact, anyone who can do this in class will be awarded an A for Physics 10. Scout's Honor. When doing the same the next semester, a student, Rodney Vasquez, came to the front of the room and succeeded in the stunt. Whoa! His brother who took my class the previous semester alerted Rodney, who then wore oversize ski boots. Rodney got an A in Physics 10 (which he would have earned anyway). The class was ecstatic. Being in my class became like being in a family.

My favorite antic is "Check Your Neighbor." After hearing an idea, we learn best if we then talk about it. That's why I directed my students to talk with one another In class—not chitter chatter, but answering well-chosen questions. I'd boldly state, "If you understand what I'm saying—if you really do—then check with your neighbor. Talk to the person sitting next to you and see if you both can answer this question." Usually the question called for a short answer, like greater, less, or no change. I sprinkled these through all my presentations. Facing an auditorium of students I considered myself as a sort of musical conductor getting the best from an orchestra. Success came with practice. What if many wouldn't participate? Eventually, I came upon a solution by boldy stating, "If you're sitting next to a neighbor who won't talk to you, maybe that person doesn't like you. Remember mom and dad told you years ago that not everybody will like you. Maybe your neighbor is one of them." Whoops! It worked, big time. Nobody wants to be thought of as a grunge. It helped to sometimes walk to and fro with a conspicuously watchful eye—but always with a smile, or better, a grin.

To maintain a family-like atmosphere in my classroom I often used the word *yum* in my lectures. I explained that yum referred to something intellectually tasty. And sometimes *yum yum*, which meant delicious. In place of asking "how come?" I shortened it to "H.C.?" a sort of verbal shortcut my students came to enjoy. Use of these terms, sparingly, added nicely to the family flavor of Physics 10.

It might be generally assumed that excelling at teaching is a common goal of teachers. Not always. I remember one of my frequent visits to the faculty coffee room during my early days of teaching when a colleague left the room saying, "It's time to toss some artificial pearls before some genuine swine," which got a lot of laughs. He was only kidding, I hoped. Did he really care about taking steps to improve his teaching?

I'm reminded of an editor of *Scientific American* who was looking for a new science columnist and scheduled a meeting to hear of my teaching ideas. I arranged a quiet corner in the college cafeteria and excitedly told him of some of my newly learned teaching tricks of the trade that could help teachers improve their instruction. On and on I went, when the editor reached across the table and grabbed my wrist, looked at me eye to eye, and in a serious tone asked what made me think that teachers actually cared about improving their instruction? He saw me as naïve, which I'll confess to. Teachers have a job. They are paid for doing that job. Whether they're successful or unsuccessful in their classes in no way affects their remuneration. He went on to suggest that most teachers value colleague recognition much more than student appreciation.

The editor's assertion was borne out by utter disinterest when the CCSF video department offered to supply a technician to video a teacher's performance in their classroom. The agreement was that the classroom video would be given to the instructor to do whatever with it, a private matter. Except for myself and one other instructor, there were no takers. I found this perplexing. By contrast, this service is very much valued in sports to critique and improve performance. Every athlete places great interest in performance improvement. It defines who they are. They appreciate any chance to refine their highly valued skills. Likewise with actors who give considerable attention to their stage performance. But sadly, this interest didn't extend to classroom teaching. Improving one's teaching performance was not every teacher's wish. This has always bothered me.

I think that making the first day of class an awesome experience for students applies to *any* class. Whatever the course level, what's left over after all the material has been forgotten is the *flavor of the course*. Was the course intellectually tasty, even delightful, or was it an obstacle to surmount? For whatever course, the flavor ought to be tasty, each in its own way. For one thing, it's important to distinguish the tools of physics (which I skip in conceptual physics) and physics itself. These tools, such as scientific notation and significant figures are essential for students continuing in science, but they don't taste good when they're useless. Sadly, they can poison a course for nonscience students.

We learn best from teachers who know their subject well, explain it beautifully, and most of all—who care about us.

There's an unfortunate reality for teaching the course for nonscience majors. When a teacher of engineering and physics is asked to teach such a course, how do they make the switch? Too often they simply teach their usual course, but watered down, and going easy on problem solving. If they cover the tools that physics majors need, however lightly, the course becomes the abhorred "killer course" that threatens the GPAs of nonscience students. The content of the course is seen as irrelevant to their educational objectives. They confront an obstacle course in pursuit only of the "almighty grade."

I found a different situation, however, when I visited the Air Force Academy in Colorado Springs during 1983. Their physics department successfully combined conceptual and calculus-based physics by using two books: my *Conceptual Physics* along with *Fundamentals of Physics* by Halliday and Resnick. The two books supported the comprehension of concepts first, then the developed knowledge to solve physics problems. In short: comprehension before computation. The turf situation common to most institutions was avoided at the Academy.

I taught my share of calculus-based courses for engineers and physicists and enjoyed problem solving as much as any of my colleagues. So much so, that with coauthor Phillip Wolf of Mount San Antonio College in Southern California we wrote *Problem Solving for Conceptual Physics*. This was a clear problem-solving book of worked examples in detailed steps that paralleled most of the chapters in my textbook. Our publisher, Pearson Education, included it in the ancillary package that accompanied the textbook. I've taken pride in that I address a common student frustration—knowing how to approach the solution to a problem. There are oodles of techniques, and mine is simply to begin by writing the symbol for what the problems asks for, then follow it with an equal sign. For example, if the problem asks for the speed of an electron curving in a magnetic field, I begin with $v =$. That's a lot. Then equations pertinent to the problems are used to guide steps to a solution. It's rather mechanical, but it works. Our book became mainly a labor of love, for it didn't get much promotion. It didn't gain the popularity we had hoped for.

During my time teaching, there was another teacher in California who achieved great results in the classroom. That was Arthur Farmer, a physics teacher at Gunn High School in Palo Alto. An impressive 90 percent of students at his public high school elected physics, compared to about 7 percent nationally. He had a great reputation for sending his students on to the best universities. It helped that among his students were the sons and daughters of Stanford professors. Nevertheless, Art's students were a mix of those seeking careers in science and engineering and nonscience students.

Art's recipe for classroom success was simple and straightforward. Instead of teaching topics sequentially from the beginning of a textbook to the end, he first rushed through the book for a "quick read." Interestingly, don't we do the same when reading a manual for any topic? As the saying goes, it's sensible to survey the whole elephant before making measurements of its tail. After the quick read that gives students an overview of what they're to learn, Art then returns to the textbook for more detail. Then back to the beginning for a final polish. Art's teaching technique

Awesome physics teacher Arthur Farmer

was awesome, and it could apply to most any science course with any textbook. Why is Art Farmer's method not generally embraced? I think the answer is simple: We teach the way we've been taught. Art Farmer's and Paul Hewitt's ways of teaching were not around back then. Teaching has many bonuses, but one stands out. When Art Farmer addressed a huge audience at an NSTA convention he stated three reasons for leaving the engineering profession to become a physics teacher: JUNE, JULY, and AUGUST! No other profession is more generous when it comes to annual vacations! None!

Physics teachers, whatever their level of expertise, are the luckiest teachers on campus. They don't need to exert an extraordinary effort to make their course interesting because physics is inherently interesting. Its topics are the bedrocks of science and are as relevant to life as any other courses on campus. Whatever the physics level being taught, the opening day of the course, if not mind blowing, should at least be uplifting. Everyone in the class should celebrate the knowledge about to be learned. And the classroom atmosphere has to be positive. My success with students had to do with mutual attitude. In my classes it wasn't *me and them*. It was always *us*. Physics 10 at CCSF was *family*.

Students have been greeted with this message in the front matter of *Conceptual Physics* textbooks from the fourth edition on. I can't say it enough: Physics presented as Nature's rules is as relevant to a student's education as reading and writing.

To the Student

PHYSICS CAN BE AN ENJOYABLE EXPERIENCE --- ESPECIALLY SERIOUS PHYSICS WHEN IT'S PRESENTED IN A NON-MATHEMATICAL LANGUAGE AND IN A DOWN-TO-EARTH MANNER. I ENJOY PHYSICS AND HOPE TO CONVEY MY ENTHUSIASM FOR IT AS I'LL BE TALKING WITH YOU ABOUT IT IN THE FOLLOWING 36 CHAPTERS. IF YOU BEGIN BY DISCOVERING THE PHYSICS IN THIS BOOK, YOU'LL SOON FIND, I HOPE, THAT YOU'RE DISCOVERING THE PHYSICS THAT'S IN EVERYTHING YOU DO AND SEE.

SINCE THE PHYSICS IN THIS BOOK IS NOT TREATED AS APPLIED MATHEMATICS, YOU WON'T NEED A CALCULATOR. MOST END-OF-CHAPTER PROBLEMS ARE EXERCISES IN THINKING ABOUT THE MEANING AND IMPLICATIONS OF THE IDEAS OF PHYSICS - IN ENGLISH - NOT VIA ALGEBRAIC MANIPULATIONS AND COMPUTATIONS. BUT IF YOU HAVE A CALCULATOR ANYWAY, SAVE IT. AS YOU FIND THAT PHYSICS IS FASCINATING YOU MAY TAKE A FOLLOW-UP COURSE. THEN YOU CAN USE YOUR CALCULATOR -- WITH UNDERSTANDING.

ENJOY!

PAUL G. Hewitt

TEN

Writing Conceptual Physics

How enjoyable it was to teach Physics 10. The textbook I used was Harvey White's *Descriptive College Physics*, complemented with Taffel's *Vitalized Physics*. But I came to love the textbook *Physics for the Inquiring Mind* by British physicist Eric M. Rogers, which addressed nonscience majors and was published in 1960 by Princeton University Press. That book remains my favorite. But it is heavy, weighing more than five pounds, with a huge trim size for that time—8½ by 11 inches. My request to have the Rogers book adopted for my class was rejected by department head Art Austin. He claimed it was much too bulky and heavy for students to have to haul around. To further justify its rejection, he also found important topics not covered in the book. I would have loved teaching from the Rogers book, but such was not to be.

Instead I supplemented the textbooks with write-ups of my own. In 1967 I wrote about the sonic boom, and an explanation of how Ohm's law is compatible with electrical transformers, for there was no coverage of these topics in either book. Also, I found White's treatment of satellite motion very unsatisfactory. He said that satellites were in orbit when gravitational force on them equaled the centripetal force on them. And satellite speed could be found by equating the equations of gravity and centripetal force. My write-up took a different view; satellites are no more than high-speed projectiles that continuously fall around and around the Earth. If they didn't fall, they'd travel in straight-line paths in accord with the law of inertia. Instead, a satellite falls beneath successive straight-line paths to follow a curve that matches the curvature of Earth's surface. For near-Earth orbit, this means that satellite speed must be at least 8 km/s, which is 18,000 mi/h. This explanation of Earth satellites in a continuous state of free fall was more easily understood than equating gravitational and centripetal forces. Writing these supplements was a pleasure, and I planned to do more.

In the spring semester of 1969, I made a big decision. I told the campus bookstore manager not to order the White and Taffel books for the fall semester—I'd instead provide a fuller assortment of write-ups for the course. This was an ambitious undertaking, but I figured I'd cover Newton's laws and other laws by having almost blank pages with only the laws stated at the top. Then students in class could fill the pages with class notes. I realized that I couldn't possibly cover all the topics in a one-summer writing binge—or could I? In any event I convinced the bookstore manager that I'd supply a manuscript, complete with sketches, in time for the fall semester. The manuscript would be printed by the campus bookstore and the cost of printing would be the price of the book.

This decision to write my own book occurred when our marriage began to sour. Years earlier Millie was happy to be married to a signpainter, but unhappy being a college instructor's wife. We grew apart. In 1969 she and the kids spent the summer in Colorado, freeing me for full-time writing in the back room of our home. I employed a small team of student typists that included Lynette Fung and her cousin Brenda Fung, Carla Jannay and her mom Hilda, and Lorraine Kahn. Several drafts contributed to a finished page, for there were no computers back then. With a dark pencil I drew sketches on final pages, making them adequate for reproduction. I began chapters consecutively, nine chapters on mechanics, five chapters on properties of matter (one which, with permission, was a direct pickup of Feynman's entire chapter on atoms from his three-volume *The Feynman Lectures on Physics*), three chapters on heat, four chapters on sound, five chapters on electricity and magnetism, eight chapters on light followed by a chapter on special relativity (which I loved teaching), and nuclear physics. All in all, a yum coverage of classical physics with my own take on pet subjects. Assistants assembled the pages into a spiral-bound book. This was also the summer of the NASA Moon landing. I was so busy with writing chores that I didn't witness most of the TV coverage. Instead, I asked friends who were living in the main part of my home to call me when the Moon landing touchdown was about to occur. They did, and I saw the climax of this awesome event, as millions of others did. With increased enthusiasm and renewed purpose, I went back to my room and continued my writing tasks.

Since much of what I wrote was the same as what I spoke in class, the writing went faster than I had anticipated. I supplemented my writing with ideas from *Physics for the Inquiring Mind*, and much from another favorite book, Ken Ford's *Basic Physics*. Other sources were Theodore Ashford's *From Atoms to Stars*, Albert Baez's *The New College Physics: A Spiral Approach*, Cooper and Smith's *Elements of Physics*, and fun ideas from *UNESCO, 700 Science Experiments for Everyone*. And of course, some ideas were gleaned from the textbooks by White and Taffel. Racing through the summer, coverage was as complete as it was in my classes. To my surprise, there were no blank pages in the completed 464-page book. I had a book I could truly teach from.

In writing my book I made sure to include a favorite remark by early twentieth-century Belgium essayist Maurice Maeterlinck, "At every crossway on the road that leads to the future, each progressive spirit is opposed by a thousand men appointed to guard the past." I also found it satisfying to state in a footnote, "In your education it is not enough to be aware that other people may try to fool you; it is more important to be aware of your own tendency to fool yourself." I see both remarks central to the wellbeing of those I could reach.

What *not* to cover was very important. For me, that was kinematics, which interestingly has no laws of physics. Teaching kinematics is a joy to teachers who love demonstrating their problem-solving skills. In a word, it's fun. And the teacher is assured that math-inclined students will also find it fun. How great to solve for the velocity of a vehicle when given hints for doing so. But response from the nonmath students is "Who cares?" Whereas topics such as waves, sound, and light are taught in weeks, kinematics commonly swallows months. My coverage of kinematics was brief in my book.

My reason for including an overview of nuclear physics was to correct some misconceptions that people have about it. Nonscientists have some idea of what the acceleration due to gravity is, but they don't have a clue about radioactive decay of isotopes in the mantle and crust that keeps Earth's interior hot. I also included a chapter on special relativity mainly because of its high interest. My classes were standing room only when I announced it would be okay to invite friends for the lecture on relativity. Relativity was sure to be in the book.

I made sure to write about the distinction between knowing the names of things and understanding them. We understand many things and have labels and names for them. Many things we do not understand, and we have labels and names for things also. I wanted my students to learn what they know, and what they don't, and progress further than naming things. Knowing the name of a thing is just the beginning step to understanding it. I wanted my students to progress beyond merely naming things to understanding the relationships between things.

Often ideas come to me during the drowsy transition from being asleep and awake in the morning. Or shortly after. One morning while spending a bit of overtime on the toilet, I came up with the idea for the title of my new book. Since the book is all about physics concepts, the word *conceptual* came to mind. That was it: the book would be *Conceptual Physics*. Coincidentally two physics professors at the University of Utah, Jae R. Ballif and William E. Dibble, had just chosen the same name for their new book with the subtitle, *Matter in Motion*. Often physics books have a common title, but a distinct subtitle.

With days to go before the fall semester began and before the book was spiral bound I had one day to silkscreen the book cover. Lettering on the cover was: *Turn On To CONCEPTUAL PHYSICS—A New Introduction to Your Physical World.* Like the psychedelic font of dance posters in the Haight-Ashbury era, the words Conceptual Physics were psychedelic in style. The price of the book was $4.20. It was a San Francisco book that some called the "hippy book" on physics.

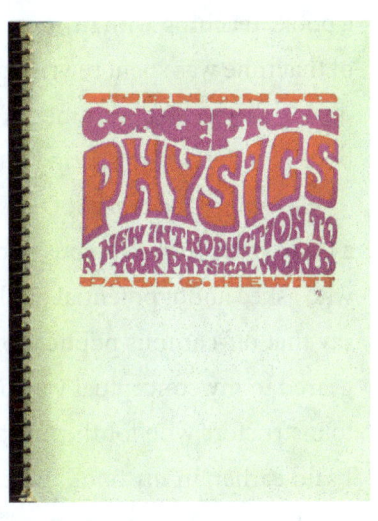

One thousand copies of the book were on time and ready for sale in the campus bookstore, enough for four daytime sections of Physics 10 and one evening section. The largest class met in the 320-seat auditorium in the Visual Arts Building in V115, the second largest class in the 240-seat smaller auditorium in the same building, and two classes in the Science Building

Lecture hall S100

in S100. All these sections totaled to more than one thousand students. I suggested that students share books, so not every student had to purchase a book. Teaching with my own book in the fall of 1969 was a joy. But much of that time was spent rewriting many pages that had been too hastily done during the eventful summer. A reprint was arranged for another thousand books for the coming spring semester.

The thousand copies of the spiral-bound one-inch-thick books occupied a lot of space in the campus bookstore. This intrigued book representatives who asked about potential publication. I advised the bookstore manager to say that off-campus publication was most unlikely because my book was geared to my conceptual way of teaching, quite local, and wouldn't appeal to instructors who for the most part taught in a problem-solving mode. As I said earlier, in my book there were NO problems. Chapter-end material was limited to only ten or so "Exercises" that were qualitative in nature, again deviating from how physics was popularly taught. Having no algebraic problems didn't mean that my book was nonmathematical. Rather it was noncomputational—a big difference.

I handled math equations in a way that became my specialty. I wrote equations on the chalkboard, and in my text, by writing symbols at different sizes to indicate relative magnitudes. For example, a central equation in the momentum chapter is the impulse-momentum relationship, **Ft = change in mv**, which I apply to a boxer being punched. Punching force **F** of the fist differs greatly even for the same impulse and same change in the fist's momentum. Moving away, **F** is small, as indicated by the large symbol *t*. Moving toward the punch, time is short, indicated by the small *t*. Aha! As the art shows, we see corresponding different magnitudes of force *F* in both cases. Similarly with the more common act of catching a fast ball. Move your hand away from the incoming ball and the force of impact is small. Move your hand toward the incoming ball and the catching force is great. These examples are common sense to students. But now they've learned a bit more detail that adds to clearer thinking. Physics, after all, is common sense with structure.

I found myself teaching *conceptualized mathematics*. Although the mathematical relationships in physics are presented throughout the book, they are without numerical values—which the students liked. I taught my students to value equations as guides to thinking. I didn't foresee publication of the book off campus, precisely because I didn't think other schools would be interested in a book without numbers. But some book reps thought that if even a few other colleges used the book, the thousand per semester at CCSF alone might justify publication. I soon found myself talking with prospective publishers.

Several book reps bought copies and sent them to their acquisition editors for review. To my knowledge, all reviews were glowing. One editor, Tom Sears of Little, Brown and Company in Boston was particularly interested in publishing the book. Other publishers also showed interest, but Tom was the most anxious to sign a contract for the book. For one thing, Little, Brown had no physics textbook. They were new to science publishing with a small number of college books and very few sales representatives.

Because I used an entire chapter on atoms from *The Feynman Lectures on Physics* published by Addison-Wesley, I felt my best course would be to publish with them, making my use of Feynman's chapter more legitimate. But Addison-Wesley was only interested if I yielded on the cartoons and let their professional staff do them. (As I later learned, using Feynman's chapter posed no problem because it was common then to pay a ten-dollar per page permission fee for such material. Since there were seven pages from the Feynman book, only a seventy dollar fee would be deducted from my royalties.) Soon several publishers expressed interest, cartoons and all. All this occurred at a time when students nationwide, and especially in California, were protesting that courses be relevant to real life. My subtitle, A New Introduction to Your Environment, resonated with students and faculty who heeded the call for relevance. The timing was perfect for *Conceptual Physics.*

Seeing a full-page ad in the San Francisco *Chronicle* helped me decide on a publisher. The ad protested the Vietnam War, and one of the signatories was Little, Brown and Company. I was very impressed, since I'd taken part in various peace marches in San Francisco. I called Tom Sears to let him know of my decision. Tom soon flew to San Francisco, and we signed a contract at his favorite San Francisco hotel, the Miyako in Japantown. That evening Tom introduced me to sushi, which would become a culinary favorite. Yum!

In the summer of 1970, I was invited to teach Physics 10 at UC Berkeley. Yet another printing was needed. This time, knowing the book would be the manuscript of a published book, I spent more time on artwork and securing permissions from authors and publishers of some of my source material. In the final stage of writing, I "borrowed" two illustrations from Ballif and Dibble, which they brought to my attention. With apologies, I paid them a small fee in line with the conventional practice.

I wrote to Eric Rogers asking permission to use renderings of many of his ideas and drawings from his insightful book. He responded with a letter congratulating me for writing a book for nonscience students and said that I was free to use any of the materials from his book, and to do what I could to bring more nonscience students into learning physics. (It was a blessing that my publisher kept a copy of that letter, for years later a law firm made inquiries about similarities between my work and that of Rogers. By then Eric Rogers died. When my publisher produced a copy of Rogers' letter, an end was put to what might have become a cash cow for the law firm.)

I asked for and received permission to draw ideas from Ken Ford's book, but when an advertisement for *Conceptual Physics* included a drawing almost identical to one in Ford's book, his publisher intervened. A $7000 settlement was made between my publisher and Ken's publisher. Ken reluctantly received half the amount, which he donated to his physics department. I say reluctant because the drawing was a rendition of one drawn by Galileo centuries ago (a block sliding friction-free on an infinite plane tangent to Earth). Since then, Ken and I have become close friends. He checks all my writing for clarity and accuracy, which has a lot to do with the continued success of *Conceptual Physics*.

With an upgraded manuscript, the book was printed off campus at Rip Off Press, a San Francisco publisher of political and new-age materials that included the comic books by Robert Crumb. This time the book wasn't spiral bound but glued, which with a lightweight cardboard cover boosted its selling price to $6.25. Ironically, the trim size of the book was almost the size of the Rogers book—which if I'd been allowed to adopt would have lessened any need to write my own book.

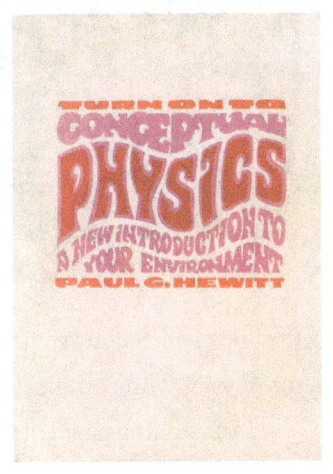

I remember being in the office of the UC Berkeley physics department when crates of the books were delivered. A couple of professors were looking through the book when I walked in to overhear one of them, "These kinds of books are a dime a dozen." Not very encouraging, I thought. But UC Berkeley students liked the book, and teaching them was a pleasure. Students were quite a bit different on average than students at CCSF. I remember a student telling me in my office of the stress he was experiencing. In high school he was the top student. In his classes at UC Berkeley were other top students from schools all around the state and beyond. Competition was fierce. For these more highly academic students I used tougher questions on my exams to get the usual bell-shaped curve that peaked at grade C.

I was delighted when the bound textbook arrived from Little, Brown in 1972. Its cover was the famous micrograph of atom locations taken by Erwin Mueller. Fellow teacher Dan St. John suggested that my name on the cover should look like a signature. Aha, I agreed, and the second printing of

Sound vibrations

the book sported my signature. By way of my book, I could share my passion for physics with countless others. I was a happy guy.

In designing pages of my first edition, I included a photograph of myself playing a piccolo near a microphone connected to an oscilloscope—and also a photo of Joe Cocker singing at the Monterey Jazz Festival. These photos played an unexpected role when I flew to Costa Rica to visit my brother Steve. At that time, hippies were unwelcome in Costa Rica. A customs inspector at the Costa Rican airport spotted me and stopped my entry. He pointed to the airplane I had just descended from and briskly told me to pick up my bags and get back on the airplane. He was firm, saying not to argue but to avoid arrest and just board that plane. I quickly removed my book from my luggage and turned to the page with the photo of me. I explained that I authored the book and effective teachers in San Francisco let their hair and beard grow long. I was not a hippy. Whew! With a brief apology the official relented and said it was okay to visit my brother who was waiting outside. I was saved by my book!

The wonder of teaching from my own book was exhilarating—all exactly on my teaching wavelength. Mission accomplished. Yet it all seemed *too* wonderful to be true. Physics 10 was described by counselors as a love affair between my class and me. It was. I loved my students and they loved me. Did I deserve this success? I thought back to my Army friend Frank Shelby saying that nobody would ever love me, and of my dear friend Ernie Brown living a lifetime of loneliness. I thought of disabled and confined people who would never experience the joy that I feel daily not only in my classroom but in standing-room-only presentations at physics conferences. Somehow my good fortune didn't seem fair—it just wasn't deserved.

Then, at home late one afternoon, in a sort of quasi sleep, I imagined myself surrounded by a circle of sages, all dressed in the clothing of biblical times. One asked, "How was it?" By this he meant, how was my blessed life. My life was beyond wonderful, for I was a loved teacher at CCSF, financially secure because of my textbook, in good health, and honored for contributions to physics education both locally and nationally. So back to the question, "How was it?" Before I could speak, another sage said, "We hope you appreciated it, and didn't blow it—we hope you didn't consider it was undeserved!" And another explained, "You drew the RED BEAN! Accept that! Relish that!" Then they explained that in a prior existence I, like others, pushed my hand into a barrel of beans, all white ones except for a scarce red one. The red bean was the guarantee of a forthcoming especially wonderful life. I had drawn the red bean. It was expected that I should graciously accept this prize of prizes, live a full and wonderful life, and be a credit to my fellow humans. I was, and am, fully blessed.

A large blessing was the top-rate editors who helped polish my writing. These began with Ken Burke in San Francisco, then to Ian Irvine, progressing to the thirteenth edition with Harry Misthos and Jessica Moro.

Another large blessing was the friendship that developed with Ken Ford. Upon the publication of *Conceptual Physics*, Ken was asked to review it for the journal *The Physics Teacher*. Interestingly, Ken had learned about my teaching when he was seated next to a young woman on a streetcar in San Francisco. When he told his seatmate that he was a physicist (strangers tend to talk in San Francisco), he was surprised and delighted when she said "Oh, I love physics." She raved about Physics 10 at CCSF.

In conducting his review, Ken showed a copy of my book to his UMass Boston students, who were using his own textbook, *Basic Physics*. Both he and his students said they liked the Hewitt book better. What Ken did next was amazing. He canceled his own book and adopted my book for the following semesters! Wow! Writing a book is an enormous effort. How many authors would drop their years of work and adopt a competing book? Ken Ford's integrity was part of my reason for dedicating two subsequent editions to him. What a great human being!

Kenneth W. Ford

ELEVEN

In Print!

The publication of *Conceptual Physics* was a life changer. For one thing, it lifted the yoke of poverty from my shoulders, the constant worry that people who are affluent in their early lives have no clue about. The twice-a-year book royalties nearly doubled my teaching salary, and eventually did more than that. Royalties meant no more borrowing to make ends meet and no more problems feeding my family. Even when my kids ordered milk at restaurants when we moved back and forth to Colorado in summers, I denied them a glass of milk, for the price of a quart of milk at any grocery store was the same. No more of that. No more dreading bills that were difficult to pay. No more poverty! This financial transition was more awesome than my words can tell. Wow!

Another great perk of being published was that my own book contained all that I wished to teach. For those who have wondered why introductory physics textbooks are so thick, the answer is simple. The publisher provides a wide swath of topics for the instructor to choose from. As thick as my favorite book was, *Physics for the Inquiring Mind by* Eric Rogers, it didn't cover topics that Art Austin wished to include in his Physics 10 course. To cover in class the topics I favored, I wrote my own. Hence the advent of *Conceptual Physics.* Having a book that covers all topics, without the excess chapters, was classroom heaven.

Little, Brown and Company placed full-page ads in two physics journals advertising *Conceptual Physics.* The ads show a slide rule with the caption, "Your students won't need a slide rule with this book!" Educators looking for a physics textbook for nonscience students were encouraged to give it a try. The slide rule was soon to become history because the calculator was making its debut. So almost four years later the ad for the second edition touted much the same message: "Your students won't need a *calculator* with this book!" An unintended inference from both ads was that *Conceptual Physics* is a nonmathematical book. This was reinforced later by the prestigious journal *Physics Today* in a complementary story reviewed by me with approval, but with a headline I hadn't seen: "In Hewitt's Teaching, Equations Take A Back Seat." As it happens, equations take a *front seat* in my book, as guides to indicate how concepts are connected. What isn't in my book are algebraic problems. *Conceptual Physics* is mathematical, but not computational. The two are very different. But confusing these concepts continues to this day.

I was surprised that my book merited a second edition. Why? Because I thought that once my message of bypassing applied math and presenting physics enjoyably became known, the "professionals" would swoop in and make my hippy book history. It took some time to realize that *I* was the professional. It stood alone. *Conceptual Physics* became well-received, big time. It went on to successive editions, each one outselling the previous edition. It was like

dancing your best on a stage and being appreciated, but with the expectation that your next dance would be even better. This process occurred over and over, including the latest thirteenth edition. In addition, I saw my influence on other physics textbooks that began including "Conceptual Questions" in their chapter-end sections. The notion of "concepts before computation" was gradually gaining acceptance. My energies in this half century have been focused on maintaining standards of accuracy and improving my physics explanations—bedrocks on which *Conceptual Physics* is built. The experience has been wonderful.

I was happy that the staff at Little, Brown and Company were diligent in the promotion of my book. During school visits, some of their representatives discovered instances of instructors continuing with their traditional books while using mine as a reference for new ideas. *Conceptual Physics* served as an interesting complement to their usual routine. What to do? My editor, Ian Irvine, offered a suggestion: Write an instructor's manual jam-packed with great demonstrations that don't appear in the textbook. This way, a teacher could use *Conceptual Physics* in their classes *and* still have plenty of good ideas not in the text. So that's what I did. I pored my imagination into creating subsequent instructor's manuals that were prized by teachers!

Writing the second edition also gave me the change to develop a "must-have" idea that I previously missed in the first edition, a lesson that I learned via numerous discussions with fellow students during my years at Lowell Tech—the value of knowing more than one side of an issue. And having a test for correctness, which I saw as being able to articulate the position of a friend whose ideas oppose yours. The most important factor is to state your friend's position to the *friend's satisfaction*! If the friend can't do likewise, your views are more likely to be correct. In the second and subsequent editions I devote almost a page to this topic to acknowledge our human tendency to shield ourselves from opposing points of view.

To bring physics to a personal level in the second edition and those that followed, I illustrated Newton's third law of motion with photos of friends and family in the act of touching, each with a caption that indicated you cannot touch without being touched. This practice contributed to the warm and friendly tone of physics.

Am I touching Lynette or is she touching me?

Is Katherine touching me, or am I touching Katherine?

Is Steve touching daughter Gretchen or is she touching him?

Is son Paul touching daughter Grace, or is she touching him?

Touching between Jennifer and daughter Io

Do I touch Lil, or does Lil touch me?

On the matter of teaching, I must say that I was fortunate to teach at a community college. High-school teaching would have involved more difficult and different challenges. This was hammered home to me one day while attending an American Association of Physics Teachers (AAPT) meeting. To other teachers, in my casual tone I remarked, "Teaching beats work!" I was comparing my prior world of factories with the classroom. Responding, a teacher bellowed back that I could say that only because I was lucky enough to teach in a community college—a very different world. Of course he was correct. High-school teachers work harder at their craft entailing enormously more daily responsibilities than I faced. Pre-college teaching is hard work. Given the lesser demands of CCSF, teaching for me was a "cup of tea."

While I was busy enjoying my early years teaching with *Conceptual Physics*, a cultural sea change, especially in music, was occurring outside my classrooms. One of the hot spots of musical entertainment at the time in San Francisco was the Winterland Ballroom, where Jimi Hendrix often played. Regrettably, I missed all Jimi's concerts thinking I'd see him later, which didn't happen. However, a student gave me a ticket to the "big one," a triple star bonanza that began with Stevie Wonder, after which the mime duo Robert Shields and Lorene Yarnell performed while band gear was being changed—and all this followed by the Rolling Stones! What a glorious afternoon! There was no shortage of quality entertainment in San Francisco.

The song "Paint It Black" by The Rolling Stones was also popular. Back then there was only one house painted black in the city, which belonged to Anton LaVey, the charismatic leader of the Church of Satan. Interestingly, his daughter was a student in my class. Fast-forward to today and you'll see many black homes and buildings in San Francisco and other cities. A testament to less-than-colorful times.

My fascination with rock concerts extended to creating a no-credit course for the evening experimental offering at CCSF. I was intrigued with the 3-D viewing of slides with double projectors. What if the slides were switched so that distant objects appeared close, I wondered. For example, a mound of soil would appear as a depression in the ground when 3-D slides were reversed. A hand pointed at you would switch to appear as the interior of a glove pointed away from you—that sort of thing. My class was a small one, but two of my students formed their own company, Optic Illusion. Later they joined the rock group Jefferson Starship and traveled worldwide, providing the light show to complement their music. I regret that the 3-D light switch idea didn't work out. However, when my brother Steve happened to visit, I wanted to share my interest in rock-concert lighting with him. At the Fillmore Auditorium, we were surprised to see how crowded it was. Attendees were not drawn to the lights, but more to the music—the British rock band Cream with Eric Clapton, and the American rock band The Byrds. What a performance!

My home in Broadmoor Village, and later in San Francisco, were a sort of mecca for physics friends. I treated those who stayed weekends to what became my routine on Sunday mornings—attending the sermons of Cecil Williams, the pastor of Glide Memorial Church. Cecil preached social justice and fair play for all and especially welcomed the down-and-out attendees and ex-felons who very much appreciated his warm acceptance. What religious beliefs I had were echoed in Cecil's sermons. When entertaining out-of-town guests, I took great pleasure in

Cecil Williams

insisting they accompany me to church on Sunday mornings. I did not tell them that the services were authentic San Francisco "be ins," rock music and all. What a joy to see my friends, especially AAPT members, finding delight in this unexpected hour-long church service. At the end of the one hour one could exit quickly, or stand in line for a Cecil hug. I always spent the few extra minutes to be hugged by Cecil. And my guests did also.

Two of my guests in my Broadmoor home were Tim and Rose Gardner, a couple that I met on a 1973 trip to Nepal. Their American traveling friend, Lori Parisi, came to pay them a visit in Broadmoor. Coincidentally, my

brother Steve with his best friend and motor-
cycle pal Milo Patterson arrived at the same
time for a few days. A highlight of this festive
occasion was the meeting not of me and Lori,
but of Milo and Lori, who connected—big time.
After Milo and Lori motorcycled in Central
America for a few months, they had a happy
small wedding in Milo's back yard in Texas. This
yum couple later built a home together in Grass
Valley, California. Ten years after they first met
at my home, they gave birth to Ryan Patterson.

Milo and Lori Patterson

Being a physics teacher took me beyond my
CCSF classrooms. A highlight was attending the
annual meetings of AAPT. I had enjoyed these
meetings from my first year of teaching. A great
benefit is the camaraderie of new friends. When
my book was released in 1972, my publisher Little, Brown and Company
presented it with fanfare at the 1972 Winter Meeting in New York City. I was
glad of this promotion, but astonished when seeing above their convention
booth a huge portrait of me! Whoa! I was not Chairman Mao or Scientology's
Ron Hubbard, whose iconic portraits defined their sects. I was an author, not
an icon. My editor promptly took the portrait down. Whew!

A similar excess of attention occurred to me when then-editor Don
Kirwin of the AAPT journal *The Physics Teacher (TPT)* recruited me in the
mid-eighties to do a monthly column that would present a full-page question illustrated with my art. I agreed, and for the first issue I decided on the
well-known question of whether a boat loaded with Styrofoam would float
as low in water as if loaded with the same weight of cast iron. The answer
would appear on the reverse page in the journal. Fine. But not fine was my
surprise of "View It by Hewitt" as the title of the piece. I don't like seeing
my name expressed so boldy. I prefer my name in the "background." So in
following issues I made the change, "Figuring Physics," with a very small

View It ———— by Hewitt

COMPARED TO AN EMPTY SHIP, WILL A SHIP LOADED WITH A CARGO OF STYROFOAM FLOAT LOWER IN WATER OR HIGHER IN WATER?

September 1986

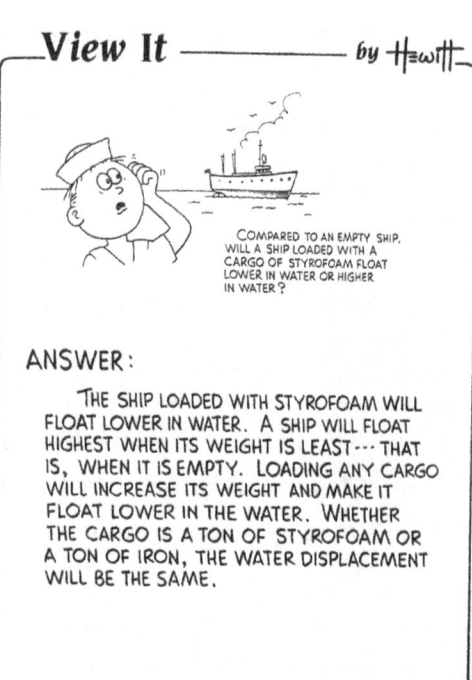

View It ———— by Hewitt

COMPARED TO AN EMPTY SHIP, WILL A SHIP LOADED WITH A CARGO OF STYROFOAM FLOAT LOWER IN WATER OR HIGHER IN WATER?

ANSWER:

THE SHIP LOADED WITH STYROFOAM WILL FLOAT LOWER IN WATER. A SHIP WILL FLOAT HIGHEST WHEN ITS WEIGHT IS LEAST --- THAT IS, WHEN IT IS EMPTY. LOADING ANY CARGO WILL INCREASE ITS WEIGHT AND MAKE IT FLOAT LOWER IN THE WATER. WHETHER THE CARGO IS A TON OF STYROFOAM OR A TON OF IRON, THE WATER DISPLACEMENT WILL BE THE SAME.

September 1986

"Hewitt Drew it!" on the lower right corner—much more sensible than just being modest. Figuring Physics continues as a monthly feature in TPT.

The first Figuring Physics issues were vertical, matching the size of the TPT pages. As teachers were more and more using computer monitors in classrooms, I realized it would be better that I matched the horizontal video screens. So I changed from vertical to horizontal formats. Another change later suggested by a new TPT editor, Karl Mamola, was to hold back on the answer until the next month's issue. This was wonderful, for I've repeatedly urged teachers to employ a satisfactory *wait time* for students seeing the question and its answer. A wait time of one month seemed quite fine, especially for questions with not-so-sure answers. Their purpose is to elicit discussion. When I posted paper copies in a glass case outside my office, waiting a week or so before posting answers, some students wished to check their answers with me. I'd suggest that instead of asking me, they should ask their friends. When the reply was that they already had, I suggested they

NEWTON'S CRADLE

Phil Physiker plays with the array of suspended steel balls in Newton's cradle. When he pulls two balls aside and releases them, two balls pop out the other side. Wondering why a single ball doesn't pop out at twice the speed, Phil figures such would violate

A. momentum conservation.
B. energy conservation.
C. both momentum and energy conservation.
D. Newton's third law.
E. More than one choice.

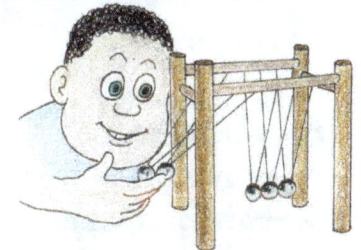

How about Newton's cradle as a classroom analogy to show sound travel in matter? Light travel thru glass?

Answer at the end of Chapter 13

make new friends. After all, didn't our moms and dads bring us up to be particular about whom we hung out with? Respect the wishes of parents.

Monthly submissions of Figuring Physics to *The Physics Teacher* were without remuneration, with the understanding that I'd rebrand them as *Next-Time Questions* (NTQ) that would be ancillaries to my textbooks. Whereas the TPT monthly questions were important to some readers for their classes, they were more important to users of my textbook as follow-ups to chapter topics. A question would be posed, then answered the next time the class met. Hence the name, *Next-Time Questions*, all arranged in order of the textbook chapters. The message to teachers was always to delay showing the answers until after a suitable *wait time*. Better learning occurs with discussions over an extended time.

Another important ancillary to the textbook is *Practicing Physics*, worksheets that could be distributed in class at appropriate times for concept development. Like the *Next-Time Questions*, they treat central physics

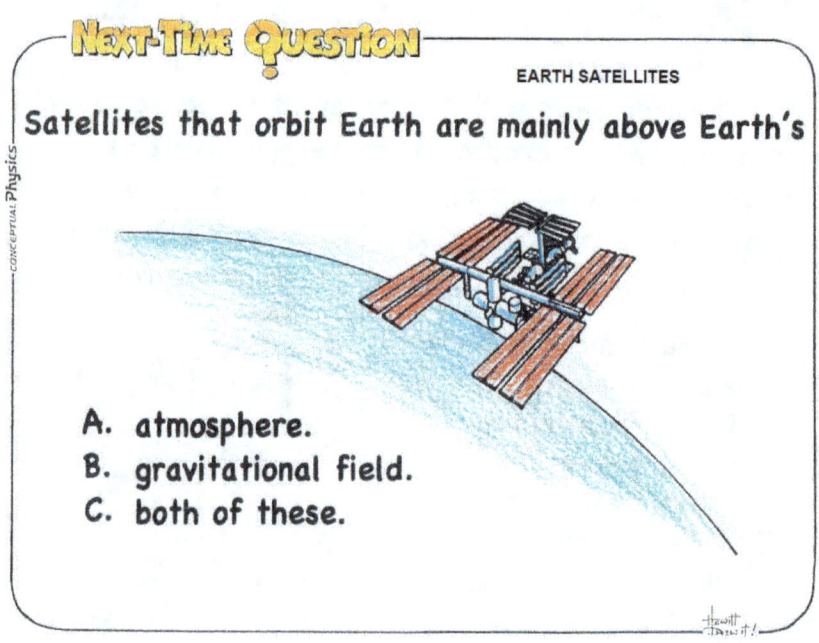

NEXT-TIME QUESTION

EARTH SATELLITES

Satellites that orbit Earth are mainly above Earth's

A. atmosphere.
B. gravitational field.
C. both of these.

Answer at the end of Chapter 13

concepts with a friendly cartoon tone. The speed of a freely falling object acquires and distance traveled are much easier to comprehend by imagining a falling boulder with a speedometer attached. The worksheet task is to fill in the missing values. More is learned when delight is a part of it.

Before the advent of Figuring Physics, my first TPT published article was a description of a simple model of "sailing-into-the-wind." The model consisted of a wooden block on wheels, into which slots were cut to hold a one-square foot aluminum sail—a flat one and a curved one. Motion in three different orientations was provided with three slots cut into the block. A hand-held fan could make the "boat" sail crosswind, along the lecture table, and amazingly, at an angle into the wind. Comparisons between flat and curved sails added to its value. Nice and simple. In 1972 I wrote another article on special relativity, which made the monthly journal cover. I would publish much more in TPT for years to come.

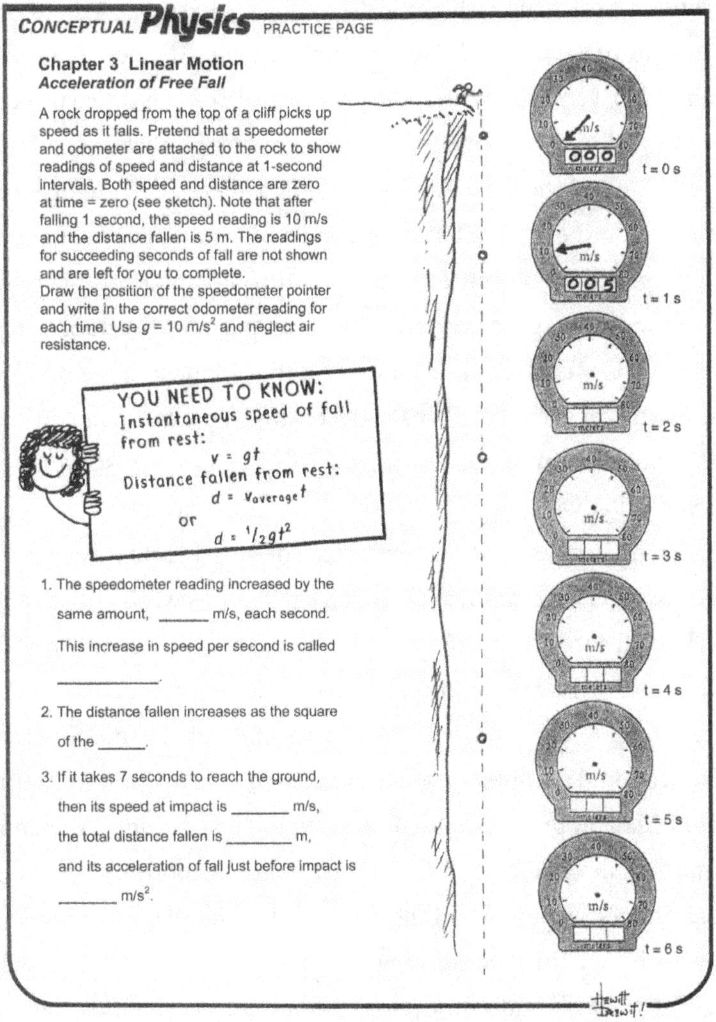

CONCEPTUAL **Physics** PRACTICE PAGE

Chapter 3 Linear Motion
Acceleration of Free Fall

A rock dropped from the top of a cliff picks up speed as it falls. Pretend that a speedometer and odometer are attached to the rock to show readings of speed and distance at 1-second intervals. Both speed and distance are zero at time = zero (see sketch). Note that after falling 1 second, the speed reading is 10 m/s and the distance fallen is 5 m. The readings for succeeding seconds of fall are not shown and are left for you to complete.

Draw the position of the speedometer pointer and write in the correct odometer reading for each time. Use $g = 10$ m/s^2 and neglect air resistance.

YOU NEED TO KNOW:
Instantaneous speed of fall from rest:
$$v = gt$$
Distance fallen from rest:
$$d = v_{average}t$$
or
$$d = \frac{1}{2}gt^2$$

$t = 0$ s

$t = 1$ s

$t = 2$ s

$t = 3$ s

$t = 4$ s

$t = 5$ s

$t = 6$ s

1. The speedometer reading increased by the same amount, _____ m/s, each second.

   This increase in speed per second is called

   _____.

2. The distance fallen increases as the square of the _____.

3. If it takes 7 seconds to reach the ground,

   then its speed at impact is _____ m/s,

   the total distance fallen is _____ m,

   and its acceleration of fall just before impact is

   _____ m/s^2.

Manuscripts for the early editions of my book were typewritten. The first two versions of *Conceptual Physics* supplied to CCSF and UC Berkeley in 1969 and 1970 were typewritten copy with art affixed. We called this camera-ready copy (CRC). But for books printed by the publisher, CRC was not required. Typewritten manuscripts could have penciled changes marked on its pages, and even Post-It notes. Art and photos were separate, with positioning indicated on the manuscript. Professional compositors used these marked-up pages to produce the clean pages of the published book. Although CRC wasn't required for the textbook manuscript, it was required for the instructor's manual. Its pages were CRC, nice and clean.

At about the time I was writing my third edition, I discovered a great technological advance in typewriters—the IBM Selectric "golf ball" typewriter. Typing was easier and faster. Instead of keys that struck a cloth ribbon to produce letters on paper, a golf-sized ball would swivel to deliver one of its 88 characters to strike the ribbon. It was a marvel. I got my Selectric in time for writing the third edition of the instructor's manual for *Conceptual Physics*. The publisher provided the "nonrepro-blue" guidelines (which don't appear on the printed page) to type camera-ready-copy. My assistant, Marissa Milan, appreciated the newest typing technology. We had a tight deadline for submission of CRC, with art glued on appropriate pages. After long daily hours, we completed and sent the manuscript to my editor on time. This occurred at a time before color photographs were in physics textbook. I took the two black-and-white photos of Marissa holding Polaroids, which appeared in the third edition of the textbook.

Marissa Milan with crossed Polaroids

Dotty Jean Rice

Helen Yan

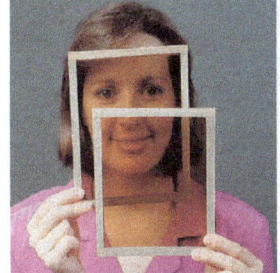

Ludmila Hewitt with a Polaroid sandwiched between two crossed Polaroids

Alexander Hewitt
and his mom Ludmila

Grown-up Alex

Grace Hewitt ponders
the steps of a lunar eclipse

Years later when color photographs became the norm for textbooks. and
Marissa was touring the world, her photo was replaced by a colored one of my

daughter-in-law, Ludmila Hewitt. The colored version appeared in successive editions. Ludmila has two lovely children, my grandchildren, moongazer Grace and skateboarder Alexander.

New photographs were introduced in successive editions of my textbook. For the third edition I took a beautiful photo of my daughter Leslie examining a model of a molecule. I call this the stardust photo as the caption explains how she and all of us are made of stardust. Leslie is looking as lovely as can be. I've repeated that photo in all my textbooks.

Classic photo of daughter Leslie with a molecular model

Another set of photographs taken before color first appeared in the seventh edition of *Conceptual Physics* was of my student Helen Yan, illustrating light absorption in a box with white interior. Helen continued her studies and years later became a "rocket scientist" in industry and a part-time physics lab instructor at CCSF, teaching the same course she took from me those years ago. How gratifying to have a student become my colleague. She repeated the same pose with the same box in color, which had graced the pages of the following editions.

Helen in black and white photo days

My grandkids' colorful balloons

Colored shadows with three lamps

It was interesting to see my book morph from no color to full color. The topic of color in physics is enchanting. I devised a lecture presentation that called for arranging three lamps, red, green, and blue, to shine on a white screen. Then I'd explain the resulting colors of yellow, cyan, and magenta that make up the shadows of someone between the lamps and the screen. The presentation is as awesome as it is entertaining. In my textbook I have a chance to show off my grandchildren in a colorful photo consisting of three layers of color plus one of black. Plus, my Lil with three colored lamps.

While I was in the midst of revising my presentations and my books, it was satisfying to follow the success of my USU friend Huey Johnson. He became California's secretary of resources in the administration of then governor Jerry Brown. When I visited him at his Sacramento office, his secretary told me that Huey instructed her in busy times to say he was out—unless it was "Dan." That's what we called each other then, which related to the *Tortilla Flat* episode from college days. I got a kick out of that. In 1978 when Jerry was searching for new blood to help his administration, Huey

The amazing Huey Johnson

told Jerry that I had integrity, something that is all too often lacking in politics. I was summoned to Sacramento. To make a long story short, I'll being just say that in the midst of a textbook revision and thinking that my success in the classroom might not ensure the same success in politics, I declined. I felt sorry to hear from the team that many candidates for the position were available, but not for the right reasons. Others were into attaining positions of power and prestige—opposites of me. Jerry needed leaders with integrity. This is reminiscent of Richard Feynman's wish for the reader at the end of his wonderful book *Surely You're Joking, Mr. Feynman!*—That whatever profession one chooses doesn't compromise one's personal integrity.

I was very pleased that one such public official who maintained his integrity was Huey. His love of Aldo Leopold's *Sand County Almanac* flavored his life. After graduating from USU in 1963 he became the western regional director and then president of the Nature Conservancy, and he went on to found the Trust for Public Land. After serving under Jerry Brown, he created the Resource Renewal Institute—while along the way winning the United Nations Environment Sasakawa Prize for outstanding contributions to the environment, and many other awards and medals. I was proud to be his friend.

CCSF in the early years was tuition free, many students needed help with other expenses. The school provided some financial aid, although often not enough to completely meet everyone's needs. Some students, especially those carrying a full load with scant time for part-time work, were hindered by financial problems.

I stepped up to the plate and helped a bit with my Albert Einstein Memorial Scholarship, an arrangement I made with the CCSF Scholarship Office. How could I in good conscience ask my students to buy a book that provided a royalty on sales without giving something in return? In a spirit of fairness, each semester I deposited funds into the scholarship kitty. The amount I deposited matched the royalties from CCSF sales of my book. Recipients received an average thousand or so dollars, plus the honor of citing a scholarship on their college record. Many students were quite happy about this.

In 1981 I combined teaching with student field trips. There were many local places to visit, such as the Linear Accelerator at Stanford, the Sunnyvale wind tunnels at NASA, the model of the San Francisco Bay in Sausalito, the top structure of the famed Pyramid Building, and so on. Plenty of good physics underscored all these sites. Knowing that the college would provide a bus, I created a new course, Physics 11, which would accommodate thirty or so students. My prerequisite to keep the enrollment low, was a student must have received an A grade in Physics 10. When I proposed the course to the administration for approval it was met with a big no-no. Why? Because the course would be *elitist*! Only the elite would be allowed—not good. I countered that only the elite are admitted to Caltech, MIT, and other prestigious institutions, which doesn't hurt the feelings of lower performing students. So long as it's not abused, elite can be a good thing. Physics 11 was approved as a once per week course. It's the only course I ever taught where all students earned an A.

One of my outstanding students in Physics 11 was Jeremy, a friendly, bright fellow who expressed hopes of becoming a high-school science teacher. I lost touch with Jeremy for a couple of years or so until one day I boarded a local bus and was surprised to see him as the bus driver. After a brief warm greeting he told me that he was granted a full scholarship to Stanford University, but reluctantly dropped out because he "couldn't keep up with" the Stanford students. Uh-oh, the classic mismatch! If Jeremy had been given a scholarship to San Francisco State University he'd likely have graduated and realized his dream of being a teacher. Matching one's abilities with those of the average student body of a school depends very much on the school one attends. I've always wondered if I were admitted to MIT instead of LTI, would I, like Jeremy, be a dropout?

In 1982, when my book was in its fourth edition, I was honored with AAPT's Millikan Award for creative teaching of physics. I was becoming a physics icon. Not everybody in the AAPT offices was happy with all this. Some likely thought, "Who is this upstart cartoonist without a PhD getting all this attention while demoting the mathematics of our cherished profession?"

Regrettably, this was the tone of the recipient of the same award one year later! I see this confusion as a failure to distinguish the goals of an introductory course from goals of courses that prepare physicists and engineers. My endeavors to clarify this distinction have been an ongoing effort. Yes, I de-emphasize computations in my *introductory* course—but certainly not in the follow-up courses. Math was not demoted. Success is sometimes a mixed bag.

In my success at teaching, I was careful not to become "full of myself." This danger of taking oneself too seriously was highlighted in one of CCSF's darkest days. That was the murder of Dudley Yasuda, a colleague in the psychology department. Dudley took himself too seriously. He saw himself as more than a teacher. He was a long-haired guru who dressed in robes, with a colorful headband across his forehead. He didn't teach conventional psychology, but instead taught his own spiritual view of the world. On the opening day of his classes, he purposely espoused far-fetched ideas that turned many students off. They dropped the class, leaving behind a band of "Dudley believers." Dudley claimed that there were no accidents, everything was preordained, and any personal problems were to be accepted as fate. For example, an asthmatic having breathing difficulties was advised to reject Western medicine and work through a karma that related to a previous life. Dudley was especially hard on one of his problematic students, so much that the disturbed student "had enough" and walked into Dudley's office and shot him to death. A clear message: In teaching, "don't take yourself too seriously," a message I've heeded to this day.

Mac Richardson

This occurred in April 1982 when Mac Richardson and Lillian "Lil" Lee were my teaching assistants. Mac was in Dudley's class in the semester when the shooting occurred. Mac emphatically told me that

Lillian Lee

Dudley was way out of order by needlessly tormenting an obviously disturbed student to death. In this case, his own. It was clear that Dudley abused his role as a teacher. Mac's sorrow was for the student.

Being humble came into focus when I met a group of high-school administrators and physics teachers at a publisher-sponsored presentation. I facetiously quipped to the teachers that they might see me as a threat to their cushy teaching positions. After all, they taught only the few best and brightest of the high school students, and they also enjoyed their premium status of being regarded as the smartest teachers in their school. So here was Hewitt, urging physics teachers to abandon their priesthood status and step down from their small classes to teach physics to the masses, ninth-graders, dirty fingernails and all. This kidding around produced fierce stares. For it was true. Not everybody was in favor of "physics for everybody."

My response was that physics for all would not dampen the present math-based physics courses. Conceptual physics would *add*, not subtract. I turn to my motto: "If a learner's first course in physics is delightful, the rigor of the second course will be welcomed and better understood." A problem with this defense is that few high schools have space for more than one physics course. As mentioned in Chapter Nine, even my Physics 10 course at CCSF was not accepted as physics credit for students transferring to four-year colleges or universities. Turf is a problem—often a big problem.

Then there's what to teach—another story. Mechanics, properties of matter, heat, sound and light, electricity and magnetism are usually taught. Topics often missing are *radioactivity* and *nuclear physics*, due to being relegated to the end of a physics course when class time runs low. This is unfortunate because these topics need attention in a time of public misunderstanding of anything "nuclear," and at a time when sources of energy are paramount. A roadblock that a teacher faces is the public perception associated with the horrors of nuclear bombs and accidents like that at Chernobyl. Nuclear power remains an emotional issue. We can pose the analogy of electricity's beginnings, which also engendered great fear. People fear most what they don't understand. The lesson to teach is that with more information the fears of yesterday need not apply to today. So it is with nukes. After the 2011 nuclear meltdown at Fukushima in Japan, a message sent to the world was "no nukes," when it more sensibly should have been "don't put nukes near locations having a history of earthquakes and tsunamis."

Know nukes!

In discussions of socially controversial issues, we should teach the *relative merits* that surround the issue. I like the idea of a "trade-off," a comparison of with and without. A good example is nuclear power. The issue is not whether nukes are safe, but whether their dangers are greater or less than the dangers of supplying the same electric power in other ways. *Trade-off* thinking should be valued by students. We ponder the role nukes will play in the future. How will nukes compare with other renewable power sources? For now, we teach that going nuclear isn't a long-term solution, but a reasoned compromise.

To appreciate the role of *time* in my presentation, I often handled an answer to a classroom question thusly: I'd write in big letters the date of the current year on the chalkboard. For example, in the 1960s I predicted that television sets would evolve from bulky cathode ray tubes to ones that could hang on the wall like a framed picture. Why was this not the case *now*? I'd write in big letters on the chalkboard "1967" or whatever the year was then. Pointing to the year, I'd remark that the reason we don't have such and such is because we're living in a time when such and such has yet to happen. It takes time for progress in every area. Why don't we have power via nuclear fusion? Why was this not that? Again, I'd answer these types of questions by writing the year on the board, emphasizing that we are living in today's time. From tomorrow's view, isn't today, we will live at a time before the advent of this or that. I'd stress that tomorrow isn't today and to have patience that tomorrow will be better than today. I must say that was easier to say before the turn of the century than after. That message would be more problematic today.

A turning point in my career began when former student David Vasquez needed a thesis project for his master's degree at Chico State University. He saw as a worthwhile topic an explanation of my success in the classroom. He

Sound technician David Milne and producer Craig Dawson

asked me if I'd mind if he brought a television camera into my class to record my presentations. That was fine with me, even though the effort expanded to three cameras and a sound system. He said this could be done without interrupting the classes before and after mine in S100. It was also fine with the CCSF administration. So it happened. David and Craig Dawson along with David Milne on the sound system, videoed the whole semester-long class. The result was the series *Conceptual Physics Alive!—The San Francisco Years.* David got

Videographer
David Vasquez

Video cameras in S100

his master's thesis and Addison-Wesley got permission to use the televised lectures as an ancillary to the newly published high-school version of my textbook. The video package was mildly successful. More important, it opened the door five years later to a greater effort at the University of Hawaii (UH Manao).

Whether a course in physics is regarded as an uplifting experience or as drudgery is important beyond school campuses. How the public perceives physics was evidenced by taxi drivers who drove AAPT participants to and from professional meetings. When a passenger mentioned that they taught physics, a typical response by the driver was, "I hated physics in school!" To a large degree this was true for many. For many years, physics was viewed as the "killer course," one that would lower your GPA. Sadly, this widespread impression may have played a role in the 1993 cancellation by Congress of the partially finished national Superconducting Super Collider in Texas. The result of that nonsupport was Europeans rushing to build the famed supercollider at CERN in Europe. There's a serious downside to public distaste for physics.

TWELVE

Exploratorium Evenings

$\mathbf{M}$y happiest times teaching took place in the years spanning the 1980s and 1990s. Much of the pleasure was presenting my physics on Wednesday evenings at the San Francisco Exploratorium, arguably the finest hands-on science museum on the planet. Arrangements were made that Exploratorium students would earn college credit from CCSF. Much of my joy was getting to know the Exploratorium's founder, Frank Oppenheimer, who was always agreeable to guest-lecture appearances in

Exploratorium founder
Frank Oppenheimer

class, especially when the topic was musical sounds and he could play his selection of flutes.

Frank was the younger brother of J. Robert Oppenheimer, the eminent physicist who oversaw the creation of the nuclear bomb. In a disgraceful period of American history, however, extensive political witch hunts conducted by the House Un-American Activities Committee (HUAC) targeted many scientists, including both Oppenheimers, and derailed numerous careers. Frank lost his university teaching job and alternative job offers mysteriously vanished, apparently due to pressure from the FBI. He and his wife Jackie decided to move to a cabin near Pagosa Springs, Colorado, that they had bought for a future summer home. The cabin had come with a lot of land, which they turned into a cattle ranch. But even here, Frank was hounded by the FBI, who also warned his neighbors that he might be doing something, anything, that threatened the security of the United States. But Frank was a good neighbor, and no one paid much attention.

In 1957 he was hired to teach science at the local high school, where he delighted in inspiring his students. Science fair judges in Denver were puzzled by the great entries from Pagosa Springs students, the children mainly of ranch hands. After ten years in Pagosa Springs, Frank was hired by the University of Colorado. While there, he developed a "library of experiments"

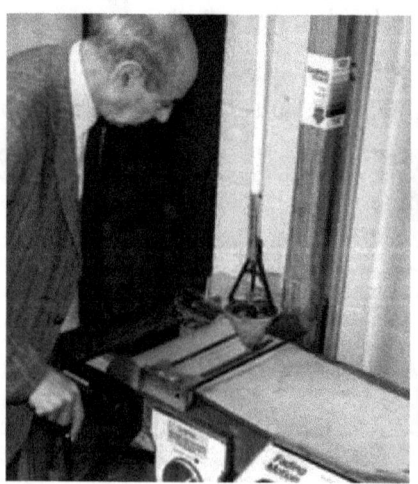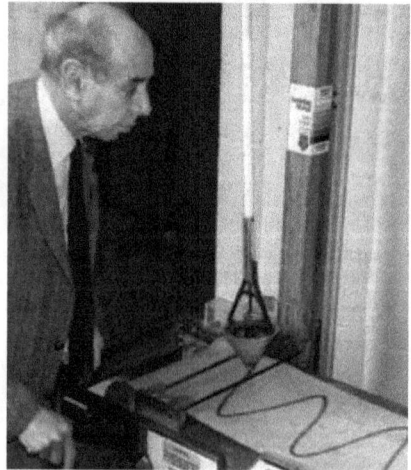

One of Frank's first exhibits

for the physics lab. Frank came to realize that he wanted to share something like this library with people outside of an academic setting. After visiting science museums in Europe, he and Jackie moved to San Francisco where he created the Exploratorium. Opening in 1969, it was filled with interactive exhibits that helped people understand our world. As Frank explained, "by trying to understand the natural world around us, we gain confidence in our ability to determine whom to trust and what to believe about other matters as well. Without this confidence, our decisions about social, political, and economic matters are inevitably based entirely on the most appealing lie that someone else dishes out to us. Our appreciation of the noticings and discoveries of both scientists and artists therefore serves, not only to delight us, but also to help us make more satisfactory and valid decisions and to find better solutions for our individual and societal problems."

Teaching in Frank's shadow was awesome. I loved the man, as did others fortunate enough to know him. I recall overhearing him in a conversation with some distinguished visiting scientists who suggested the universe was meaningless. Frank's response was that he preferred to think that even in a meaningless universe there are *islands of meaning*. In my mind, one of those islands was the very science museum he created—the Exploratorium.

Frank often sat in on my classes. He was pleased when physics lessons had social implications. I felt particularly lucky that he visited when my topic was phase changes of water. The lesson involved distribution curves, which are important in assessing not only molecular speeds in liquids and gases, but variations of peoples' shoe sizes, heights, and intelligence.

I drew a bell-shaped curve on the board while asking my class to think about the great distribution of water molecule speeds in a saucer of water. Some molecules are slow, some are fast, while most move at in-between speeds. Interestingly, the slowest at any moment may be hit by others to quickly become the fastest. Molecular motion is helter-skelter, and is nicely depicted via a bell-shaped curve. The left side of the curve depicts slower-moving molecules with low kinetic energy (KE), while the right side is of faster-moving ones with high KE. Students easily understand that the faster-moving molecules with the greatest kinetic energy are the ones most likely to break free of the water. This is evaporation. The students already had learned that water temperature is directly proportional to the average KE of its molecules. What does the escape of high-energy molecules say about the average KE of molecules remaining in the water? Average speed will be less, which further tells us that temperature will be lower. The temperature of water decreases when evaporation occurs. This explains why evaporation is a cooling process.

Looking at the distribution curve, I then remarked about its generality. For one thing, it applies to student test scores. In my daytime classes I post exam scores in a display case near my office. The sketch shows a photo of the actual distribution curve for one of my classes. Note how it approximates a bell shape.

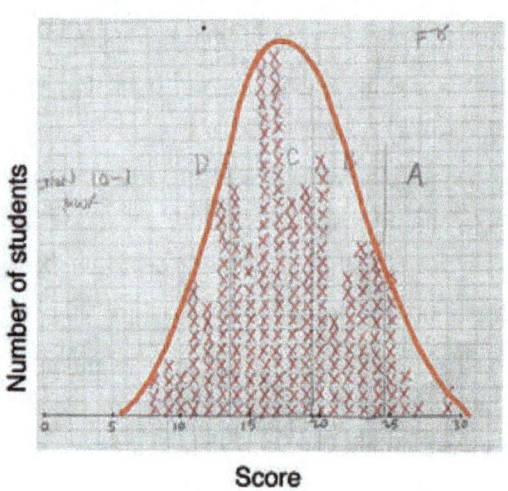

Distribution curves of exam scores

I had two daytime sections of Physics 10, Groups A and B. Curve A of my 10 a.m. class shows a higher average than the Curve B of my 1 p.m. class. The two groups superposed show a large area (blue cross-hatched) where the curves overlap.

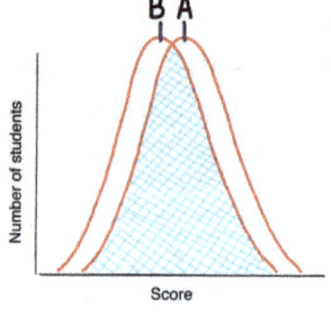

Overlap of wide curves

This overlap is important—very important. To stress its importance, I ask my students to imagine that a teaching assistant enters the room to give me a note. After the assistant leaves, I mention that the assistant is from Group A. Then I ask, "Can anybody make an

assessment of my assistant's exam-scoring ability?" It should be clear that the answer is no.
The assistant's score likely falls in the large blue
region area where the curves overlap. That's why
"no" is the emphatic answer. Hence this message: We cannot judge an individual person's
performance based on the group the person
belongs to. If the curves of the two groups

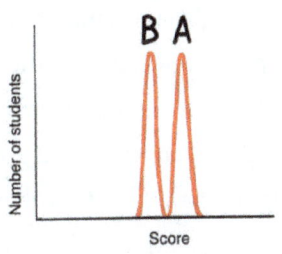

No overlap of narrow curves

didn't overlap, but were narrow spikes instead of wide bell curves, that would
be a different story. The wideness of the curves attests to the wide variety of
performances in a group. Hence the difference between narrow-mindedness
and broad-mindedness. Frank was pleased with this interpretation of distribution curves. He agreed that prejudice between groups of people would be
lessened if more people thought broadly rather than narrowly. Physics is on
the side of reasoning.

Then we have curves that represent intelligence—sometimes controversial. I want to point out two thoughts about them. One is the claim that the
IQ of people is fixed. My personal experience tells me otherwise. When I was in
the Army we all took IQ tests and our scores were posted beside our names. Ten
years later when I was in graduate school, my high Army score was seven points
higher. IQ is *not* something that is fixed. This is important in my teaching—
that I'm lifting smarts as well as imparting information. The second thought

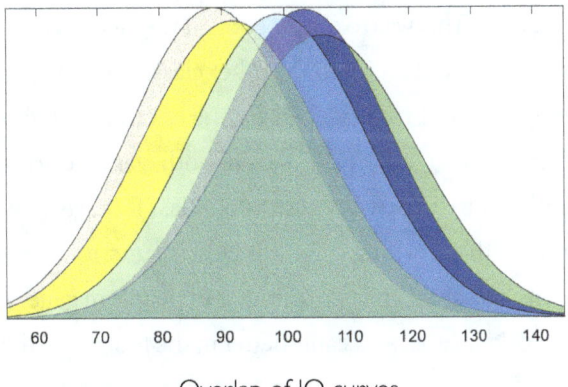

Overlap of IQ curves

I wish to share is the significance of the overlap in IQ curves. Consider two cups of coffee: one a little bit warmer than the other. Are all the molecules in the warmer cup moving faster than in the colder cup? Absolutely not! Likewise, consider two groups of people: one with an average IQ a little bit higher than that of another, such as a morning physics class compared to an afternoon physics class. Can you say all students from the morning section are smarter? Absolutely not! There may well be some geniuses within the afternoon class. Thus it is I find it disturbing when people try to generalize the intelligence of any individual merely based upon the group they belong to, be that a physics class or an ethnicity. Understanding the broad overlap is key. More importantly, a person's worth is much more than is indicated by an IQ score. So let's just get along.

A major part of getting along is the wonder of love. The Beatles had it right when they sang, "all you need is love." We all value it, for it's the best of being alive. What matters to each of us is how much

> "But in the end
> The love you take
> Is equal to the love you make"
>
> BEATLES

we love. For the love we give is the love we get. So when someone laments they aren't being loved, the advice is to start loving others. Love is a two-way street. It is highly likely that the love they give will be the love they receive. Ask any pet dog or cat.

There's more of interest with higher performance in the morning classes. When my classroom was first videotaped, cameras recorded both the morning and afternoon classes. This was to offset "goofs" I might make in one class that could be replaced with footage from the other class. An interesting discovery was that not only were test results lower for the afternoon class, my teaching was also less impressive. I assessed videos of both classes during the editing phase and found we rarely used any afternoon-class footage in the final product. My teaching in the morning class was clearly superior. I later found this was in accord with research: The performance of people in general, whether at home or in the workplace, is noticeably better in the morning (although not in the very early morning when people are drowsy). I wonder how exam scores would change if classes were switched. Timing is truly important.

In 1990 Huey Johnson introduced me to a neighbor on Telegraph Hill, Gary Zukav, the author of the popular book about metaphysics, *The Dancing Wu Li Masters*. Interestingly, Gary wrote most of this book during afternoons at the nearby Caffe Trieste. The book is a good read, even though some of the physics was quite unconventional—less than good science. Gary told me that he hoped to convince the reader that we live in a conscious universe with religious purpose. He supported his views on the cracks within quantum physics. After many discussions, I discovered his main point. Gary said that he couldn't imagine anything more horrible than the notion that we live in an unconscious universe—that when we die we're eaten by worms with no hope whatever of continuance. He saw his mission as giving people hope of something better, to avert the ultimate horror of death. I told him that I didn't see things that way and that I had no problem with the notion of no afterlife. Before we were born, we experienced nothing. Was this so bad? And after death, we're back to where we came from—to nothingness. If the time after death is like the time before birth, what's the problem? Is nothingness so fearful? Can't that be accepted? Where's the horror? Besides, as said by Frank Oppenheimer, we can build islands of meaning in a meaningless universe.

Toasting good physics
with Gary Zukav

Celebrating my fiftieth birthday party

I invited Gary to visit my Exploratorium class. After a brief introduction, Gary agreed to answer student's questions. One philosophical question that I can't remember was difficult to answer. Rather than saying he didn't know, Gary did what gurus seem to do—he smiled a bit, then rolled his eyes toward the ceiling, and remained quiet. Then he asked if there were any more questions, as if to ask, any more *sensible* questions. He reminded me of Dudley Yasuda. I didn't invite Gary to speak at future classes. Before Gary moved from San Francisco, he attended my small fiftieth birthday party. I last saw him on the Oprah Winfrey show where Oprah held his hand and thanked him for giving hope to people.

People in all professions make mistakes. In the teaching profession this is especially so. I am one of them. I remember one evening at the Exploratorium when the topic was light and color. That was when two or three women strangers appeared in my class after the break. It was common for visitors to drop in like that. I had three colored pieces of card, red, green, and blue. Showing the red card, I explained to the class that the ink resonated with the same frequency as red, and hence, produced the red that we see. The women apparently had come to check on the quality of content of this "charismatic teacher." They left in a huff. Was my physics wrong? Yes, it was! In my classrooms over the years, I had been teaching incorrect physics. The card was red because it *reflects* red. Lights of other colors resonated and were absorbed in the ink, allowing red to be reflected. So if it were said that Hewitt is more style than substance, in this case that was right. I made sure my physics of color was correct in my textbook, and in later classes. Again, the beauty of teaching is that continual improvement is inevitable. Learning is an ongoing process, and that's especially true when teaching. We keep getting better at it—as long as our heart is in it.

The many exhibits outside the Exploratorium's classroom theater were a teacher's heaven. Having exhibits to play with that complemented the lesson made the experience immeasurably better than the same class in a common classroom. Fifteen-minute breaks were time very well spent on the Exploratorium floor. Gabriel Espinda, my assistant and an Exploratorium explainer, was happy to guide students to the relevant exhibits. For example, if the topic of the week was electricity and magnetism, Gabe would direct attention to the simplest exhibit, Ohm's law, that shows how current relates to both voltage and circuit resistance. By a careful sequence of exhibits he could transition to Faraday's law of electromagnetic induction. Students get a feel for what otherwise couldn't be visualized. Gabe excelled at his lucid explanations, which gave me time to address individual student concerns.

On alternate weeks many students accepted my offer of free drinks at nearby bar, Liverpool Lil's. I explained how I felt a bit guilty collecting royalties for the textbook they purchased. Royalties for day classes at CCSF went into funding the Albert Einstein Memorial Scholarship that I established as mentioned earlier. Exploratorium royalties? We'd drink them at Liverpool Lil's! Most of my Exploratorium students were of legal drinking age. A few were physics teachers from surrounding schools seeking teaching tips. Teaching at the Exploratorium was a joyful experience for all.

Frank Oppenheimer was the most-loved teacher in my experience. He was also a chain smoker and died of lung cancer at age seventy-two in 1985. Another great teacher was Albert Baez, the co-inventor of the X-ray reflection microscope with Stanford physicist Paul Kirkpatrick. Al's specialty had been coherent light, before the advent of the laser. Al gladly accepted my requests to teach my Exploratorium class when the topic was light-wave interference and

Albert Baez

its corresponding colors—like the vivid colors of soap bubbles. In addition to being a skilled lecturer, he authored books on physics. A popular one was *The New College Physics: A Spiral Approach*. It covered topics lightly, then in more detail, with each spiral adding to a more thorough explanation of covered concepts. Much of his teaching was in Middle Eastern countries, particularly in peacetime Baghdad, Iraq. His spiral approach to physics was international in scope, with the support of UNESCO. It also influenced my own writing. Al's fame as a physicist became eclipsed by his folksinger daughter Joan's fame. At public gatherings he was used to being asked if he was related to Joan. He told me how delighted he was when at a conference someone looked at his name badge and asked if he worked on coherent light with Kirkpatrick at Stanford. How nice it was that the fame he was credited for was directly his.

As a sidelight, Ken Ford tells me that when he was in Cambridge, Massachusetts, in 1958, he met seventeen-year-old Joan, who was his Vespa motor-scooter instructor. When they exchanged names she told Ken her name and spelled it. Ken said, "That's an unusual name. Any relation to the physicist Albert Baez?" "Yes," she said, "he's my father." So, at that time in Ken's universe, Albert was the more famous Baez.

How delightful it was for me to teach with these giants, Frank and Al. When Al turned eighty and I was living in Hawaii, I was invited to his San Francisco birthday party. Without hesitation I flew to San Francisco and when attending his party asked him a favor: Would he attend *my* party when I became eighty? I'll always remember his response, looking up at me with a twinkle in his eye, "Don't count on it!" He assumed he wouldn't be around for that. As luck would have it, that almost could have happened. With a bit of good luck Al may have lived to my eightieth birthday.

The cherry on the Exploratorium cake was a banquet in my honor and the 2002 Lifetime Achievement Award. In a nearly tearful acceptance, I told of my earlier encounter with Frank when he gave me a personal tour of the illusions exhibits. "He ended the tour," I said, "with an experiment I want to share with you. Everybody hold your arms up a bit and look at the sizes of your hands. Both look the same size, right? Now bring one hand half as close to your eyes. Which hand looks bigger? Do some of you see the close hand as being about twice as big? It's more than that! Close one eye and look again, but this time slightly move the closer hand so it almost overlaps the farther one. Aha! Can you see the closer hand is twice as tall and twice as wide? We see the *inverse-square law* in action. The closer hand appears four times as large as the farther hand. Quite amazing!" Then Frank uttered the punch line: How many other illusions do we have about our surroundings that aren't so easily checked? Think about that! What if most of what we see,

Frank's influence on my lecture

upon closer examination, isn't so? My story of Frank and apparent hand size has been one of many that I've shared with my classes. Frank was an incredible influence on me.

Family and friends took my Exploratorium class. One of my "adopted family" was Jenny Jones, the daughter of BJ, my "man in Thailand." See on Chapter 16. On an earlier visit to Thailand, I took a photo of a much younger Jenny sitting on her Uncle Ben's lap, pondering the fate of a chickie. The full-page photo with a stimulating message in cartoon fashion was a hit with readers of *Conceptual Physics*. It was especially appreciated by her dad who displayed it in his hotel in Pattaya. I was pleased to see a grown-up Jenny in my class.

Among the relatives were Corine Jones, the little Pocatello niece in USU days, who became a physical therapist. Then niece Nancy Hewitt from Massachusetts, whose brief stay in San Francisco was followed by some hippy years in La Honda, and then to Ireland, where she attained some fame as a songwriter and singer—now a Shade Landscape consultant. Another relative was

| Corine Jones | Nancy Hewitt | Robert Baruffaldi with Leslie |

nephew Robert Baruffaldi, also from Massachusetts, who went on to become a meteorologist.

Friendships made while teaching at the Exploratorium included impressive senior scientist Paul Doherty, who to my joy and his usual big smile ducked

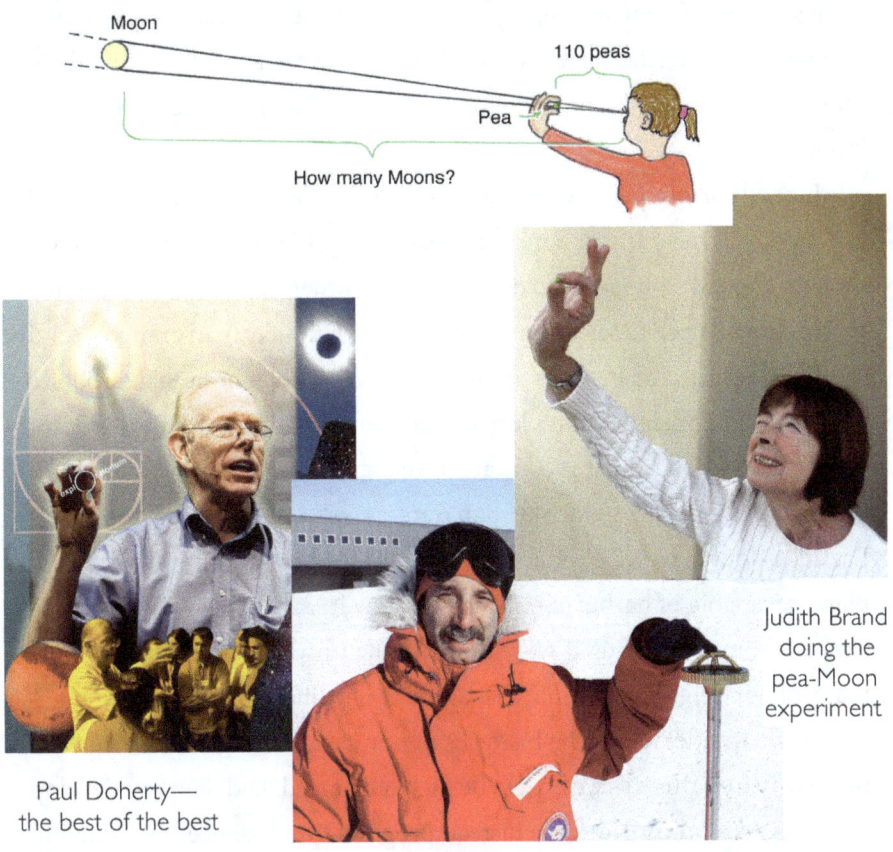

Paul Doherty—
the best of the best

Judith Brand
doing the
pea-Moon
experiment

Ron Hipschman on assignment at the South Pole

in and out of my classes. Every summer he'd teach an Exploratorium work-shop for physics teachers. I learned a lot of physics from him. He was loved by thousands. Sadly, Paul Doherty passed away at age sixty-nine in 2017. Two other Exploratorium staff members are Judith Brand and Ron Hipschman. Judith was a science writer and editor for Exploratorium publications and often worked with Paul on a variety of projects. More recently she became the developmental editor of the thirteenth edition of *Conceptual Physics* textbook, and nicely contributed edits to this memoir. Judith's photo shows her hold-ing a pea at a distance from her eye that barely eclipses the Moon. Aha! The finding that 110 peas would fit between the pea and her eye tells her that 110 Moons would fit between her eye and the Moon. Yum physics! Or actually, yum geometry. Scientist Ron, a great lecturer at the Exploratorium, has the distinction of being the longest serving Exploratorium staff member. His photo shows him at Earth's South Pole. Both Judith and Ron remain close personal friends.

Like all curious people, I wonder a lot. Frank's assertion that there could be "islands of meaning in a meaningless universe" was a sample of what I thought about. I wonder about civilizations on other planets. At the outset of my teaching, I used to say there are more stars in the sky than grains of sand on all the deserts and beaches in the world. Wow! Then noticeable wobbles of some stars indicated that planets were out there too. And more recent confirmations tell us that most stars have planetary systems, which means there are more *planets* in the sky than stars, as well as more grains of sand on Earth! A double wow! To add to this, recent studies indicate "Goldilocks" planets, capable of harboring life as we know it. A triple wow! We've progressed from wondering if we're alone in the universe to wondering how many of our neighbors are like us—all this in my lifetime!

So we wonder about the histories of our galactic neighbors, especially those who may think as we do. Or, be more advanced. Did they first develop religions, as we did? Did some instead progress from science? I used to kid with my religious friends that we should send copies of our Bible to our planetary neighbors to save their souls. More seriously, I ponder what we'll eventually learn from these neighbors. I wish I had insights into these questions. Sadly, I don't. Nor, as far as I know, does anyone else. Still, we wonder.

The Exploratorium that both Frank and I knew was located at the Palace of Fine Arts, on the northwestern edge of San Francisco near Golden Gate Bridge. In 1969 Frank rented the space from the city for one dollar per year. I began teaching there in 1980 until my retirement in 1998. In 2013 the Exploratorium moved to a larger space at Pier 15 on San Francisco's historic Embarcadero, not very far from the landmark Ferry Building. This was unsettling for me, but when I visited it, I still saw the hand of Frank. I was happy with this iteration of the Exploratorium, and I like to think that Frank would be as well. It remains, in my thinking, the world's finest hands-on science museum.

A common but greatly valued message given by most parents to kids growing up is to take care with who they hang out with. The quality of one's life, after all, has much to do with the company one keeps. Being with such people as Frank Oppenheimer, Albert Baez, and others cited in this chapter certainly added quality galore to mine. I remain grateful to them.

Frank Oppenheimer

Old man writing his memoir

THIRTEEN

Writing Other Conceptual Textbooks

One of the delights of authoring a textbook is getting letters of thanks from teachers far and wide. Most of these appreciative teachers are users of *Conceptual Physics* in high schools. Although my focus on successive editions of my book was community colleges and universities, my publisher promoted it to high schools as well. The story of how and why I was prompted to write a high-school version of my book began with the honor I received in 1982, the AAPT's prestigious Robert A. Millikan Medal for notable contributions to the teaching of physics.

Receiving the award meant giving a presentation. I prepared my speech like a madman, with the coaching of CCSF colleague Neil Fleishon. The previous year's awardee, Al Bartlett, used overhead transparencies to guide his talk. I did the same, but I sprinkled my transparencies with cartoons and different-sized symbols for equations. I titled my paper, "The Missing Essential in Teaching: A Conceptual Understanding of Physics." I began my talk by acknowledging what we all know, that mastery of math and familiarization with the tools for doing physics are essential for engineering and physics students. I mentioned another essential too commonly missed in traditional physics instruction—*conceptual understanding*. My opinion was that all physics classes should nurture the ability to conjure a mental image of a physical interaction, process, or concept, and to be able to describe it verbally, and to distinguish the concept from others that are closely related. First conceptualize, then compute! Let's look at the whole elephant before we begin to measure its tail. I delivered all that I cherished in my teaching to a receptive audience that honored me with a standing ovation. I was elated.

The Millikan Medal was akin to the Silver Medal earned decades earlier in boxing. I had two silver medals, both highly valued. I liked to kid with my friends that I much more valued the boxing medal. Why? Because it took more courage to enter a ring and duke it out with a fighter you'd never seen before than to be a good teacher. And adding to this courage, as I mentioned in Chapter Two, was revealing my skinny body to an auditorium of people. In contrast, I was fully clothed for the Millikan talk.

In discussing my award with friends, it became evident that my conceptual-teaching message should logically be directed toward high schools. The sooner that students study physics, the stronger their foundation will be for the other sciences. My quest then turned to writing a high-school adaption of my book. After gaining permission from Little, Brown and Company, I sought a high-school publisher. I thought this would be a shoo-in, given the large numbers of college adoptions. Not so! Why? Publishers weren't interested in a new high-school physics textbook because physics, of all the sciences, drew the lowest enrollments. A physical science textbook would be better supported, but not a physics book. Aha, as luck would have it, publisher Wayne Oyler, at Addison-Wesley, had wished to acquire my initial book in 1970 when he worked for a different company. At that time, Wayne was overseas when the book was signed with Little, Brown and Company. He felt he had lost out. Upon hearing of my high-school request, he instructed his acquisitions editor, Shelly Moore, to sign a Hewitt contract. Whew! That was done and I gladly spent the next couple of years writing the book that Wayne had wished for.

One of the teachers who convinced me to write the high-school version of *Conceptual Physics* was Fred Myers, who used my college second-edition textbook to teach freshmen at the prestigious Choate Rosemary Hall in Connecticut. At that time, and still today, students in high schools in other parts of the world generally took physics either before or at the same time as they took chemistry and biology. At that time in the United States, however, the sequence was quite different: biology first, then chemistry, and then maybe physics in the junior or senior year. Courses in this order was established in the late 1800s when only a few students completed a 12-year education. These were very different times. Learning personal hygiene was deemed most important, so biology was taught first. The electron was yet to be discovered, and chemistry was mostly a course in mixing chemicals. So chemistry was taught in the second year. More mature students with

Fred Myers

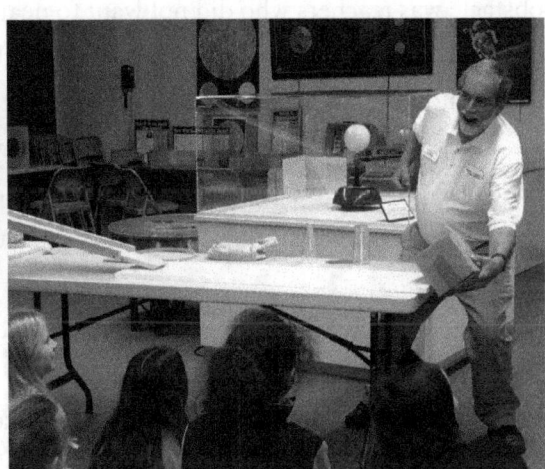

Fred engaging students

higher math skills took the problem-solving course, physics, in the senior year. To Fred this sequence wasn't right. It seemed upside down—especially because physics is generally acknowledged to be the foundation of chemistry, and chemistry the foundation of biology. Fred was one of the few in the United States to invert the traditional sequence and teach physics in the ninth grade—before chemistry and biology. Enthused by his success with high-school freshmen at Choate, he earned a second master's degree and moved to Farmington High School, a public school in Connecticut, and successfully implemented a physics-then-chemistry-then biology curriculum.

I was enormously impressed with Fred's published article advocating "The Right-Side-Up Science Sequence," which became well known as "physics first." Wow! This was a compelling reason to create my high-school version of *Conceptual Physics* as "physics first," aimed at ninth-grade students. Thank you Fred! Later he was recognized by the 1990 Presidential Award for Excellence in Science Teaching, presented to him by then President George W. Bush. This was the highest of several other prestigious awards that honored Fred.

Fred tells me that getting acceptance for physics first was like being a mosquito pushing against a truck. He was disappointed to find that the main obstacle was teachers who did not want to deal with "average kids." Physics teachers enjoy teaching the best and the brightest without having to manage difficult student behavior that other teachers endure. Some also like their status as being "the smartest teacher in school." I have to wonder: how many teachers of average smarts choose high-school physics teaching to gain that almost priesthood status? There must be some, but I hope not too many.

I credit Fred for introducing me to the term "killer course." By this he meant the overly difficult physics course that kills a student's GPA. In 2006 he moved on to administrative positions where he continued the physics-chemistry-biology sequence. With this new order Fred emphasized the notion of delight over challenge as part of the growing physics-first movement. Fred and his wife Bonny remain close personal friends.

Where to begin writing the high-school textbook turned out to be interesting. A favorite relative I enjoyed being with was cousin Bud Baruffaldi, especially on summer visits at his family farm in Landaff, New Hampshire. How nice to begin writing the high-school version of *Conceptual Physics* in this secluded location. I was still getting used to computers replacing typewriters and arranged borrowing a portable computer from

Bud Baruffaldi

my publisher, Addison-Wesley, who had an office in nearby Lynnfield in Massachusetts. Writing went smoothly in this quiet environment. Bud had previously treated me to a tour of local maple syrup farms. I was astonished to learn that fifty gallons of maple sap is required to produce one gallon of

maple syrup. Wow, fifty to one. That meant repeatedly boiling an initial fifty gallons of sap and ending up with a single gallon of syrup. It reminded me of my task of boiling down an enormous amount of physics to create my textbooks. It was common to see very thick introductory physics textbooks, which in my mind were "information overload." As with my college book, my mission was to continue creating a thinner book with only the syrup of physics—the correct amount for respectable introductory physics textbooks.

A distinguishing difference between my high-school and college books is that the high-school book addresses a mix of both science and nonscience students. Science students value math. Hence, there's a bit more emphasis on math in the high-school book. I focus on the equations of physics, for both science and nonscience students as guides to thinking. I express

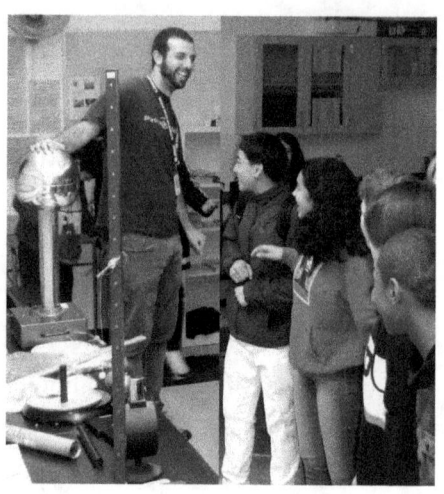

$$I = \frac{V}{R} \; ; \; I = \frac{V}{R} \; ; \; I = \frac{V}{R}$$

magnitudes of quantities with different-sized symbols. My teacher friend, Elan Lavie, reminds me of how clear thinking guided by equations is especially valued when teaching electricity

Elan Lavie electrifies his students

and magnetism (E&M). Without equations, visualizing the relationships between concepts can be difficult to grasp. This isn't the case so much in mechanics, where equations present a formal way of thinking about what's already familiar. Everybody knows, for example, that a greater push means more change in motion, and that anything moving faster has more energy. But E&M is completely mysterious until physics swoops in and lifts the curtain! The equations of E&M are essential to understanding what's going on. As the sketch shows, the relationship between quantities in Ohm's law are clearly indicated in its equation.

I want to acknowledge comic strips that graced the pages of my high-school book. This relates to Ryan, the son of Milo and Lori Patterson. I watched Ryan's growth from childhood and have "adopted" him as my nephew. Because he was very bright and an Eagle Scout in the Boy Scouts of America, I hoped he'd embrace a life of science, especially physics. Lori enrolled him in a local Catholic school. My high-school version of *Conceptual Physics* was in print and my hopes were that his high-school teacher would adopt

Ryan Patterson

my book. I made an appointment with him to show my new high-school textbook. The teacher leafed through its pages commenting on the cartoon figures and the full-page comic strips. He was not impressed. He didn't say so, but his expression seemed to convey, "Our kids don't need cartoons . . . they can do the math." I feared that response. Ryan took the physics course and I

was saddened to learn that the teacher chose the widely used college textbook for physics and engineering majors written by Halliday and Resnick—the same book I taught from in Utah and CCSF! Ryan struggled and passed

the course with an A, but he told me he never wanted to confront physics again. He hated the course. Many teachers pride themselves on rigor in their courses. For Ryan's teacher, cartoons and comic strips didn't indicate the proper amount of rigor. What he didn't notice in his quick review, was that

the comic strips made clear some of the toughest-to-understand concepts in physics. Because they were greatly appreciated by students, the comic strips remain a big source of pride to me. Ryan didn't enter a science profession, but went on to become a successful lawyer.

With the growing popularity of both my college and high-school books, the attendance at my sessions at AAPT meetings grew, often with standing room only that extended into hallways. To accommodate the overflow, scrambles often occurred in switching to larger rooms. I was happy that Addison-Wesley's Wayne Oyler arranged for hour-long promotional workshops in the largest available rooms, where I could acquaint teachers with the wide swath of ancillaries that accompanied both the high-school and college textbooks. This publisher support was gratifying.

The success of *Conceptual Physics* prompted the writing of *Conceptual Physical Science* (CPS). Several friends and colleagues had suggested this over the years. It was important to write a textbook on physical science because physics courses were unpopular. Physics was not a course many schools wanted. A physical science course had a better reputation and could more easily make its way into classrooms. I could teach the physics I wanted via physical science textbook. A conceptual approach to physics, as well as chemistry and earth science, *would* reach the most important audience—the school teachers who would relay this kinder approach to science to countless children in their classrooms.

In writing physical science, my wish was for a family team of authors—to make the project a family affair. My coauthors would write chapters on chemistry and earth science with some astronomy, and I'd edit them with a conceptual flow to produce a textbook that would read as if written by a single author.

My sister Marjorie's son, nephew John Suchocki, had just completed his PhD in chemistry and through spirited discussions I was delighted to find he was on my conceptual wavelength. Assuming he had the same talent for writing as his mom, I chose John to do the chemistry. It was easy to convince him to join me in Hawaii, teach at a local college or university, and write

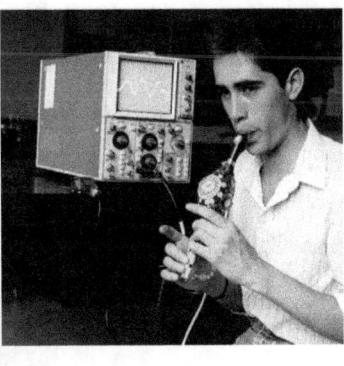

James in light and sound

chemistry chapters. He expressed his take on science: "If there's one thing we aim for, it's not so much the meaning of life, but a life of meaning. All pathways that lead us to greater meaning are to be valued." John was my guy!

For years I had wanted to inspire my son James to join me in writing, especially after discovering through a friend who belonged to Mensa that James found solving Mensa problems an easy task. Of my children, James was the brainiest one. But sad to say, he was also the most stubborn one. All my kids resisted following my path of higher education. Son Paul was born with learning disabilities. He graduated high school, which was good enough for him. Daughter Leslie was endowed with great looks and a charming personality. She was busily enjoying a rich social life and at least attended CCSF to

Leslie

Paul with his cousin John Ray, and brother James

attain an AA degree. I wished to point my bright son James toward a college direction and presented him with an offer I thought he couldn't refuse: Get a college education in any science and I'd guarantee he become a coauthor for a yet-to-be written physical science book. Then, whatever else he chose to do in his working life, every six months he'd receive royalties—maybe even large ones.

James replied no, he wasn't ready for that, and he'd rather sidestep college and pursue other avenues in life. Leslie, aware of this amazing offer, asked, "If James is too stupid to accept your offer, can it apply to me?" My answer was yes. So Leslie took an academic path and earned

Dynamic trio: me, John and Leslie

a BS in geology from San Francisco State University. She was on the way to complement her geology degree with an MS in geography, but instead choose elementary education. Nevertheless, Leslie kept her part of the bargain and authored Earth-science chapters, which with chemistry chapters by her cousin John Suchocki, resulted in the fine *Conceptual Physical Science* textbooks. Both Leslie and John turned out to be excellent writers. Hooray to that!

Along with all conceptual textbooks was a lab manual. Finding a reliable author for a lab manual was a bit rocky at first, but then we discovered Dean Baird at a local AAPT meeting. Dean is exceptional in several ways.

Among his several teaching awards at Rio Americano High School in Sacramento, his top one was the Presidential Award for Excellence in Mathematics and Science Teaching, which was given during the Obama administration. Dean traveled to the nation's capital for the ceremony and was greeted by Vice President Joe Biden, who pinned the medal of accomplishment on

Dean Baird

Dean, since Obama was temporarily out of the country. Dean's passion for physics is matched by his love of photography. Many of the photos in my books were taken by Dean, who never missed a publishing deadline. Like nephew John, my kind of guy!

For two years I guest taught a conceptual physics course at the University of Hawaii at Manoa, in Honolulu. During my stay a dedicated teacher and AAPT member Pauline Chinn rented her two-bedroom unit on the thirty-second floor of an elegant condo, Mott-Smith Laniloa to me. How could I pass this up. I was in a tropical heaven.

After my teaching stint at UH Manao, I turned my teaching to Hilo, the Big Island. Continuing with my writing chores I purchased a nearby condominium unit at Mauna Loa Shores (MLS), which became our "book factory." MLS is nicely situated on a stunning beach surrounded by swaying coconut trees, a site that previously included the residence of Hawaiian royalty. Unit

Bedroom view of a double rainbow in Hilo

706 was on the top floor and soon was doubled when I purchased a next-door unit and joined the two. Its beautiful Hawaiian furniture and great view of the ocean were attractive to frequent guests. I was pleased that several couples enjoyed their honeymoons there. When I resumed teaching in San Francisco, my vacated MLS units were open to guests galore. One of them was the Exploratorium's Paul Doherty, who started a guest book that remained part of the Aloha magic. I was amazed that Paul had seen all the attractive sights of the Big Island in one week, places that took me more than a year to discover. With time the guests were less frequent, and a downstairs neighbor who loved the top floor with its high ceilings and who cared for guests in my absence purchased the units in 2010. I made an even swap for the price of her unit. It was a happy transfer.

Back to my son James. Ironically, he found employment doing what I had done years earlier. He printed T-shirts in a small silk screen firm in Salida. On February 29, 1988, James was killed in an automobile accident. Shocked beyond belief, Leslie and I traveled quickly to Salida. We learned that James had been promoted at work and his friends celebrated his promotion with Irish whisky and beer chasers. A friend needed a ride across town, and James, although quite intoxicated, drove him home via a safe back road. Returning on the same road that had new lane markings he was unaware of, he collided head-on with an approaching car. This occurred in front of the home of our family physician, Doctor Mehos who ran from his home to the scene. James died quickly in the doctor's arms. The sudden death of James was more devastating to me and my family than anything else had been in our lives. The worst had happened. Getting over this loss was difficult for all, but I was somewhat helped by an expression James frequently used: "Shit happens." I felt some consolation by extending his saying: "Shit did happen, but big time." Uncomplicated. It wasn't easy to return and face my classes at CCSF and the Exploratorium. Somewhat choked up, I managed, and gave my students a homework assignment: Tell

Millie with first grandson

Manuel as a tot

Manuel illustrating a catenary   Celebrity chef Manuel those you love that you love them. Never mind that they know this. TELL THEM, again, if need be.

James had a son, Manuel, who was my first grandson and has been a joy. This was especially so when he lived with me in Hawaii, where we enjoyed many evenings watching Beavis and Butt-Head on television together. Manuel graduated from high school in Hawaii and later from CCSF, to became a prize-winning celebrity chef. The photo of Manuel holding the chain with the photo-shopped background of the Gateway Arch of St. Louis is a favorite, showing how both the arch and the chain form catenaries. Manual has delighted both Millie and me. Dark clouds do have silver linings.

While Leslie and I were working on the second edition of CPS in San Francisco, a dear friend Lil happened to visit. Seeing us trying to solve paging problems, though not wishing to barge in, she asked if she could help. We were delighted to get her very useful input—for page layout as well as in catching typos. This was the woman I would be working with on later editions and other publications. No doubt about it, to say the least, Lil was and is an impressive editorial assistant.

CIS author team: me, Suzanne Lyons, John Suchocki, and Jennifer Yeh

Conceptual Physical Science went through six editions. Later we included biology to create *Conceptual Integrated Science* (CIS), which called for a writing team of strong credentials. For geology we recruited my former Addison-Wesley editor Suzanne Lyons, with a BS in Physics from UC Berkeley, MS in Earth Science at California State University in Sacramento, plus a MA in Science Education from Stanford. Being very familiar with our conceptual approach, Suzanne was the right choice. Once aboard, Suzanne recruited biologist Jennifer Yeh, who had graduated summa cum laude in physics and astronomy from Harvard University, then earned a PhD in biology from the University of Texas at Austin, then acquired teaching experience and authored various scientific papers. In addition to John's and Jennifer's PhDs, and Suzanne's graduate degrees, I had a highly-credentialed and talented first-rate author team.

Our textbooks were aimed at college students. But we became concerned when we learned that many high schools, and even some middle schools, were adopting them. Uh-oh, the reading level for older students might not be good for younger ones. Our goal has always been to get more students liking or loving science and to counter the anti-science sentiments prevalent in parts of the general population. That meant producing books that match student learning abilities. So we wrote *Conceptual Physical Science—Explorations*, for a younger audience. We did the same with *Conceptual Integrated Science—Explorations*. These Exploration textbooks seemed to work well for middle and high school students. Students loved the books, as did the author team.

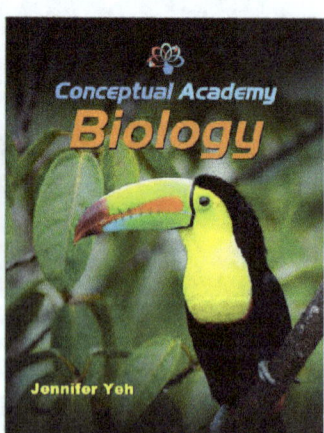

But in the financial downturn at the turn of the century, our publisher no longer promoted the Exploration books. Fortunately, author John came to the rescue and secured rights to them so that they could be recast for the Conceptual Academy audience. The author team has since rewritten the Exploration textbooks, taken care to avoid the *information overload* common to other textbooks, while adhering to state requirements. In keeping with delight in learning, these books feature mascots. There is Sneezlee the conure, Pio the penguin, and Manny the mouse. A new mascot is Tookie the toucan. To highlight equations as expressions of nature's rules, the thirteenth edition of my book features cartoon-mascot Manny the Mouse. We see Manny affirming one of several central laws of physics—which indeed are awesome!

An awesome rule of Nature: $F\Delta t = \Delta mv$

In addition to authoring textbooks, I began writing physics for a general audience. These are called trade books, which are common in bookstores. My first was published by Ron Pullins and his wife Leslie of *Focus Publishing*. Its title was *Conceptual Physics for Parents and Teachers*, which was sold mainly in bookstores. It didn't sell well. Pearson secured a later version when a Pearson official asked for a title change that would be, as he said, more "sexy." What was meant by more sexy went over my head. I retitled the

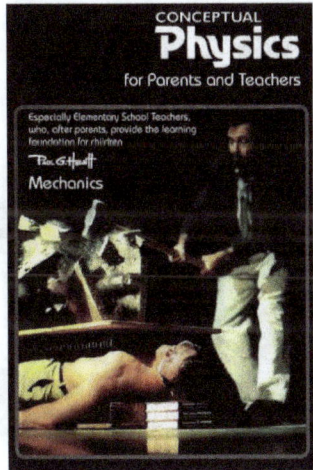

book, *Touch This!—Conceptual Physics for Everyone*. The cover photo of two little girls, Debbie and Natalie, the daughters of friends Hideko and Herman Limogan, and a boy, Genichiro Kakada, between them at Golden Gate Park in San Francisco displaying we can't touch without being touched, as my way of expressing Newton's third law. The photographer is David Vasquez, and the cover designer is Lil. I retitled the book, *Touch This! Conceptual Physics for Everyone*. It's a fine photo and a nice book. But as poor book sales revealed, it became nothing to write home about.

A nice benefit to *Touch This!* book was becoming acquainted with a reader, Bruce Novak, who contacted me via email in July 2011. Many readers made good suggestions over the years, but the many remarkable ones in Bruce's letter were most impressive. I considered them worthy of a gold star! I responded by asking if he'd be available as a reviewer to my future writings and he agreed. He made significant contributions to the twelfth edition of *Conceptual Physics*. He also gave valued feedback to the fifth and sixth editions of *Conceptual Physical Science*. John, Leslie, and I were so pleased with

his input to the sixth edition that in addition to honoring lab author Dean Baird, we dedicated the edition to Bruce Novak. Rather interestingly, after hurricane Sandy knocked down two trees in Bruce's back yard in late October 2012, he used most of a bonus check that I sent him

Linda and Bruce Novak   The Hewitt Tree   Bruce creates a color spectrum

to purchase a ten-foot tall "October glory red maple," known since as the "Hewitt Tree," to replace his fallen trees.

How satisfying that my books have been so well received. I had a very special personal hope for *Conceptual Physics*. I wished it to become as important a teaching resource for secondary school teachers as *The Feynman Lectures On Physics* were to university professors. How much of a resource it became, I don't know, but I'm content knowing that it's been used by many.

Richard Feynman remains a physics hero of mine. He puzzled some people when he often said he didn't know anything. Of course he meant that what he *did* know is closer to nothing than what he *can* know. He was referring to realms beyond his experience. For example, What was his take on humanity's place in the universe? Are there galactic neighbors "out there" who think as we do? Or more advanced? Feynman offered no answers to such questions. Closer to home, my friend Huey Johnson used to suggest that when searching for a solution to a problem, find where a similar problem has occurred and was solved. We apply this common sense when we train our telescopes to other world "out there." What can be learned from extraterrestrials yet to be contacted? Assuming we are not alone in the cosmos, what are the commonalities of successful worlds out there, and can they apply to us? How I'd like to be alive when answers occur.

FIGURING PHYSICS

Phil Physiker plays with the array of suspended steel balls in Newton's cradle. When he pulls two balls aside and releases them, two balls pop out the other side. Wondering why a single ball doesn't pop out at twice the speed, Phil figures such would violate
A. momentum conservation. B. energy conservation C. both momentum and energy conservation.
D. Newton's third law. E. More than one choice.

Answer: B

Although momentum conservation would not be violated [$(2m)v = m(2v)$], energy conservation would be [$2(\frac{1}{2}mv^2) \neq \frac{1}{2}m(2v)^2$]. One ball popping out twice as fast would have twice the energy of the two released balls—not going to happen! Experiments attest that incoming balls always come to rest and an equal number of outgoing balls leave the opposite side. However, this isn't required by the conservation laws—but by the forces exchanged between the balls. For example, both conservation laws are satisfied for the hypothetical case of one released ball hitting at v and rebounding at $v/3$, causing two balls to pop out at equal speeds $2v/3$. Check the math! Conservation laws alone can't always determine the results of an interaction, but importantly, interactions that *do* occur must satisfy these laws. Only Choice B is correct.

LIGHT

The maneuvers of colliding balls are *consistent* with conservation laws, but not *explained* by them.

How nice when one demo apparatus illustrates more than one physics concept. Bumping balls are akin to bumping molecules that transmit sound. Light too!

For light travel in glass; mechanical energy flowing from ball to ball in Newton's cradle is akin to photons flowing from glass atom to glass atom in a gulp-burb sequence. The photon to exit is *not* the same one that entered!

NEXT-TIME QUESTION

Satellites that orbit Earth are mainly above Earth's

A. atmosphere.
B. gravitational field.
C. both of these.

Answer: A

Did you say C, both of these? Oops, an Earth satellite is *not* above the gravitational field of Earth — nowhere near so. Earth satellites are mainly above the atmosphere and are strongly in the grip of Earth gravity, which is why they don't fly off tangentially into space. Earth's gravitational pull on the International Space Station, for example, is nearly 90% that at Earth's surface.

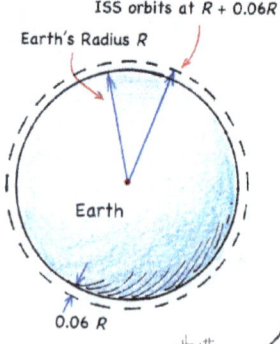

ISS orbits at $R + 0.06R$

Earth's Radius R

Earth

0.06 R

FOURTEEN

Teaching at Various Universities

My first invitation to be a guest lecturer was in 1970, to teach Physics 10 at a summer session at UC Berkeley. This was an honor. Thanks to funds for book production from Little, Brown and Company, I rented a studio apartment on Channing Way, very close to the campus, which meant I didn't have to commute daily from Broadmoor to Berkeley. My textbook for Physics 10 was the prepublication version of *Conceptual Physics* by Rip Off Press. This was a dream come true—to teach from my own textbook. This was also my last chance to make corrections and improvements before the advent of the hardcover first edition. I redrew much of the art and cleaned up many of my explanations. It was a labor of love.

One summer evening I took a walk along Bancroft Way to the corner of Telegraph Avenue, the main entrance to the UC campus. What I was about to see was unforgettable. This was the era of Free Speech and anti-Vietnam War movements. A year before, on Telegraph Avenue there had been a riot at People's Park. An aftermath of that riot was the march that was approaching me. Leaders carried a large "People's Park" banner. The march was peaceful. Most of the men in the march had long hair and beards, as I did. But I saw two shorthaired beardless men with bandanas on their heads, and they seemed out of place. Telegraph Avenue was under repair, with many loose rocks about. Horrified, I saw the two men, only them, throwing rocks from the street into store windows. Other marchers cried stop, but the throwers yelled political slogans while continuing their mayhem. Then they disappeared. Very soon the police and media people arrived. What was in the news and newspapers the next day? Something like, "People's Park protestors ravaged Telegraph Avenue in a smashing-windows spree." I had heard of how a few outsiders can delegitimize a peaceful movement. I witnessed this myself.

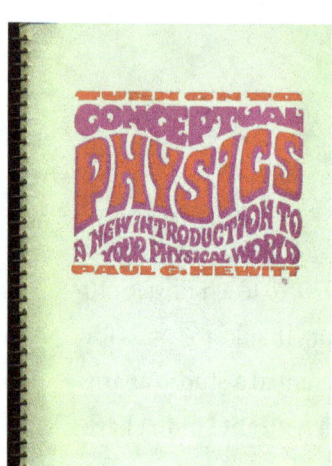

Back at CCSF in the fall of 1971 the bound book arrived. My pleasure was intensified. Any errors in the book would be corrected by the publisher in successive printings. Little, Brown and Company was generous in this respect. I announced to my CCSF class that I'd pay five dollars for each typo or other error that a student discovered. And, there were quite a few. One student earned nearly a hundred dollars by identifying wrongly hyphenated words. She simply looked down the column of sentences on each page and circled each error. It was fun and games for the class to watch for errors, and in some cases, to improve wording. This was when the bookstore price of the book was $9.95. So finding two boo-boos would pay for the book. As said, it was a fun time for all.

In 1980, ten years after the initial publication of my book, I returned to UC Berkeley and taught two quarters, this time with my fourth edition. I was fortunate to teach in the huge PS1 auditorium next to Le Conte Hall. The stage was a large one that rotated out of view of students when apparatus for lecture demonstrations was being assembled. The small team that operated the rotation of the stage did more to encourage demonstrations—they assembled and disassembled them. Wonderful! There were no shortages

of demonstration in classes taught in PS1. Another feature of the auditorium was the ease of entering and leaving at the back of the hall. One could slip in and out to eavesdrop on classes without causing a disruption.

The easy entrance to PS1 enabled my mother to enter and find a seat while my class was under way. She had never seen me teach a class and said she'd try to visit. Although PS1 was in a hard-to-find location, she asked

My mom

directions from students and made it. She witnessed her Paulie doing his thing before all these upscale university students. Tears came to my eyes, and I took a quick break from my lesson to introduce her. The class greeted my mom with applause. My favorite comedian George Carlin had quipped that life is not measured by the number of breaths we take, but by the moments that take our breath away. For me, this was one of those moments. Another such moment was when I learned that my rating by UC

Berkeley students for that semester was the highest ever recorded in the physics department. I was appreciated every bit as much by Berkeley students as by CCSF students. I could have remained there as a lecturer, but I felt more needed by the lesser-advantaged students at CCSF—my home base.

One of my awesome experiences in the PS1 auditorium occurred in a casual visit to witness Professor Owen Chamberlain teaching Physics 10. His topic was atoms—and amazingly, from my textbook! How gratifying to watch him steer students' attention to the subject of antimatter at the end of the chapter. In his closing remarks for the class, he told them that he was the discoverer of the antiproton, which earned him the 1959 Nobel Prize in physics. I was in awe.

Biking with Yoshihisa Yoshida

At Berkeley I shared an office with a Japanese guest of the university—professor Yoshihisa Yoshida. He liked my textbook and I found him photocopying its fourth-edition pages. Whoa, said I—not needed. Shy Yoshi was elated when I gave him a copy of the book. He encouraged me to visit his university in Japan, and I accepted, especially when I learned that he had memberships at clubs for dancing in Tokyo. We shared more than a love of physics—we shared a passion for dancing. A nice feature of Tokyo dancing spots is that everybody wishing to dance simply gets onto the floor and dances. Nobody had to ask a partner to dance. No fear and no embarrassing "No thank you." Everybody danced solo, and if a connection with a prospective dancing partner occurred, okay. If it didn't happen, okay. No stress. All fun.

Dancing in Japan was enjoyable, but more engaging were the physics conferences I attended with Yoshi. The most notable one was in Okinawa.

The meetings were akin to the AAPT conferences in the States, but with a big difference. At
the conclusions of such meetings there was
singing! Attendees mounted the stage one at
a time to sing a song. Any song would do, with
the help of lots of sake beforehand. I had never
done this, and chose a song I knew the words
to: The Alphabet Song—"A you're adorable, B
you're so beautiful," . . . and so on. Of course
remembering the words wasn't important at
all. Nobody cared. I appreciated this added
twist and very much enjoyed the conference.
But mainly, I was there to show my short movie
"Relativistic Time Dilation," which elicited
much more interest from the Japanese educa-
tors than it had received at a recent AAPT meeting in New York.

My turn at singing

Yoshi was an only child. He told me how in his childhood he and his
mom scoured the landscape looking for food while Tokyo was being fire-
bombed. He spoke of glowing red skies above Tokyo, at a time that preceded
the nuclear bombing of Hiroshima. His mother lived long enough to appreciate her two grandsons from Yoshi's marriage. I was fortunate to meet his
mom when I was invited to sleep over at their home. A small bedroom was
always available for my visits. From the bedroom window I could see the
construction of a small building next door. I was impressed with the corner
constructions. Instead of nails being driven to join the wooden frame, the
ends of the wood were carefully chiseled to hold together without nails. Ah,
I thought, this craftsmanship would show the quality that Japanese workers
took pride in. I was wrong. The next day plywood slabs covered what I saw
as a work of art. That was my first impression of the quality I always saw in
Japan.

Yoshi went on to please his mother, and his studies as a teen culmi-
nated in becoming a physics professor at the Sagami Women's University in

Sagamihara. Yoshi, with the help of professors from Tokyo University, translated the first edition of *Conceptual Physical Science* into Japanese. This was a book coauthored with daughter Leslie and nephew John Suchocki. When daughter Leslie visited Japan in 1982, she slept in "my room" at Yoshi's home for a month.

Enjoying physics with Al Baez

Yoshi was active in international physics conferences. One year in Mexico City he met Mexican-American physicist Albert Baez, who invited Yoshi to stop by whenever he was in the States. Al lived in Greenbrae, just north of San Francisco at the end of a walkway over the water of San Francisco Bay. His prefabricated octagonal home was a birthday gift from his folksinger daughter Joan, who had it delivered by a huge helicopter. Whenever Yoshi visited San Francisco, he included me. So we both had a new friend.

Yoshi did not trust conventional medicine and for most of his life put his health in the care of a homeopathic medicine man. When Yoshi visited mutual friend Howie Brand in Thailand, his health was in question. His medicine man told Yoshi that his problem was due to poor diet. Howie strongly advised that Yoshi see a conventional doctor, which he didn't do. This is remindful of Steve Jobs, who ignored the advice of those who loved him and put his trust in alternative treatments, which resulted in an early death. Likewise with Yoshi. After spending all his savings on a blood cure recommended by his medicine man and with advanced cancer, he finally sought a traditional doctor. But it was too late. In 2011 he died at age of seventy-two. No funds were left for his family.

Back to the USA and 1980. Immediately following guest lecturing for two quarters at UC Berkeley, I was invited to teach two quarters at the University of California at Santa Cruz (UCSC). My home base at CCSF approved of this extended absence. A big reason I got away with such favors is that I was a "good guy" at CCSF. I never had a chip on my shoulder with the physics department or the administration. I always got my grades in on time. As my brother Steve would say, I always had my ass in the seat when the bell rang. So I accepted the UCSC invitation and very much relished the experience. For one thing, the campus is situated in a beautiful redwood forest overlooking the Pacific Ocean, as stunning as a campus can be. For another, the professors were impressive, and friendly. The classes were smaller than I was accustomed to and easier to manage.

Letter grades are the traditional way of measuring academic performance. But not when I taught at UCSC. At the time, UCSC was the only campus in the UC system to use narrative evaluation instead of grades. All courses were on a pass/no pass basis. This made good sense. Sometimes grades say more about the professor than about the student. For example, we all know of "easy-A" professors. Student assessment in the pass/no pass policy cites a student's standing in the class—upper 10 percent, upper half, lower half, or whatever. Additional narrative essays about the student's performance were also required. So the pass/no pass system provides more, not less, information about each student. It's actually very nice, but not so nice in terms of the professor's workload. Much less effort is involved with letter grades. Nevertheless, I was happy with the pass/no pass policy because it seemed sensible, and it motivated me to better know my students. Years after I left UCSC, its pass/no pass system reverted to the letter-grade system. The option of pass/no pass in student assessment applies since to about a quarter of its courses.

In 1986 or so Department Chair Alex Burr invited me to be a weeklong guest professor at New Mexico State University in Las Cruces. This first classroom experience outside California was made possible by my colleagues at CCSF who substituted for me for that one week. I enjoyed the visit very much and thought of myself as a happy goldfish in a new bowl, swimming with different fish. Foremost of these was Alan Van Heuvelen, the author of the premed physics textbook we used at CCSF. We sat in on each other's classes. Part of that week was giving an address in a large auditorium. I talked with the students, interspersed with several "check-your-neighbor" spots. My audience ate it up and I loved it. This wonderful week set the stage for other similar outings. And most importantly, it made me think about teaching in Hawaii for a year.

Stopovers in Hawaii had occurred regularly when I was returning from Japan and Southeast Asia. For one thing, they reduced jet lag. My friend Ron Fitzgerald from CCSF was always there to welcome me on these short stops. He had graciously provided photographs in my first edition. Without question, Hawaii is one of the most lovely places on the planet. Another friend Pauline Chinn urged me to spend more time in my Hawaii stopovers and join her in workshops.

I was already acquainted with some of the physics faculty at the University of Hawaii at Manoa in Honolulu. My textbook was used in a course similar to CCSF's Physics 10. In Hawaii the course was Physics 100. Rather than submit an application for part-time employment, I telephoned the department chair, Chuck Hayes. I didn't ask for a one-year guest lecturer position, which would have involved more than one class and a lot of paperwork, but instead asked to guest teach only one course, Physics 100, for one year. Chuck replied that the remuneration for one course would likely be less than the cost of air travel. Not a problem, I said. This is just a meaningful vacation. My request was accepted.

For this teaching adventure two teaching assistants were on hand: Ted Bradstrom and a fellow named Troy, whose last name escapes me. Both were very helpful. I was delighted to have my own office, something I hadn't experienced at CCSF. Furthermore, I taught my class in a great auditorium. I was in a new heaven. My students were as wonderful as I'd hoped for, were receptive to learning some yum physics and we all had a great time.

Professors at UH Manoa were impressive both in teaching and in research. A most notable researcher was John Learned, one of the founders of the Deep Underwater Muon and Neutrino Detector (DUMAND) Project.

His quest was to discover whether or not ultrahigh-energy neutrinos ejected by astronomical sources have mass. It was an open question at the time. His project involved arrays of submerged spherical detectors suspended by cables anchored to the ocean floor. John was a very highly respected physicist and a great host at his many well-attended home parties. His experiments, unfortunately, met huge obstacles related to unpredictable ocean swells. Years later, he and his team joined the prominent Super-Kamiokande project in Japan, which confirmed that neutrinos do indeed have mass.

Promoters of UH Manoa encouraged me to give frequent presentations to students in high schools, mainly to discuss the offerings in science enjoyed by university students at UH Manoa. In my closing remarks I'd recommend they do something that might not meet the approval of their parents—to NOT go directly to college after high-school graduation. Take a one-year "time out," a gap year. I cited my own experience of "growing up" before entering college at age 28. Taking a study break can add to one's character. If hard work is involved, it makes college life something to look forward to even more. Why would parents disapprove? Because of the high chance of not "staying on track." They are correct about this, so students would have to resolve that after a year off they'd be back on track, and with a more mature outlook in life.

A memorable classroom day in Hawaii occurred when my topic was rotational motion, which involved rotating a bucket of water at arm's length over my head. Coincidentally, this was also the day of a much-publicized Space Shuttle launch from Cape Canaveral in Florida. Students and I were chatting about the launch before class time. A few minutes into my presentation, a momentous thing happened—as I swung the bucket of water in a vertical circular path, I asked the class to compare the water at the top of the swing to astronauts who at the moment were likely overhead in the orbiting Space Shuttle. Voilà! A splendid connection—yet another insightful analogy in my bag of tricks! The water in the bucket was falling, but in a circular path. The astronauts in the Space

Shuttle also were falling, but in a much larger circular path about Earth. Yum! I have used this analogy over the years that followed. How nice that learning is a continuing process.

The next semester was a mixed bag. Technical problems at UH Manoa were being addressed in an attempt to link all the Hawaiian Islands to the TV studio at the Manoa campus. Professor Hayes knew of my experience in front of TV cameras at CCSF some years earlier and asked if I'd teach the spring semester in their TV studio. I quickly agreed, and shortly thereafter realized this provided an opportunity to upgrade my earlier CCSF videos with better cameras. New videos could augment the *Instructor's Manual* that accompanied my textbook. So I asked if I could supply master tape cartridges and own the copy of the video footage to promote my textbook. The answer was yes. After all, what the university wanted was video practice, and officially I could do whatever I wanted with my video duplicates. So, it happened. Just as I had no idea that my first edition would go on to successive editions, I similarly had no idea that my videos would be watched for decades in school classrooms far and wide. Those classroom videos, called *Conceptual Physics Alive!* are now my legacy.

All teachers have bum classes from time to time. Sad to say, the Hawaii class that was videotaped was my worst—the most nonresponsive class in my experience. And not by coincidence. Unresponsive student behavior was mainly due to a camera problem I failed to recognize. Students were intimidated by one conspicuous camera above my lecture table that pointed directly into their faces. Due to that roving camera, few students raised their hands with questions. Why didn't I realize the situation and simply disconnect the camera—or cover it with a towel? I since regret not doing that.

The following semester my course returned to the physics auditorium—and back to cheery students. How I wished these students were the ones in the videoed classroom. As said, it wasn't the students, but my failure to deal with the roving camera. My only consolation is that teachers who use the videos will feel their own classes are more responsive than Hewitt's. Hewitt doesn't inspire *all* his students. At least there's something good to feel about.

My Chicago physics teacher friend Marshall Ellenstein agreed to edit the master videotapes and he skillfully produced a series of thirty-four lessons. Marshall has been a resource from the time I met him at a Chicago AAPT meeting in 1977. A meeting of the minds resulted in his many good suggestions for both the textbook and the creation of the *Practicing Physics* booklets, and the *Next-Time*

Marshall
Ellenstein

Questions. The expenses incurred in producing the thirty-four-lesson video set were recouped by sales in the competent hands of Peter Rea of Arbor Scientific. In his catalogs he listed and highlighted VHS tapes of *Conceptual Physics Alive!*. Furthermore, Peter showed sample lessons at AAPT conventions. As the times changed, the video set became DVDs that spread my method of teaching far and wide. And soon after the era of streaming began. This led to the creation

of *Conceptual Academy* by nephew John Suchocki and me. Currently all the *Conceptual Physics Alive!* lessons are available free of charge with no advertisements to all except corporate interests. Access them on YouTube or Conceptual.Academy. Take a look. Let's spread the word that learning physics can be a pleasurable experience.

A heartwarming story about the use of my videos in classrooms comes from primo teacher Nina Brady at American Charter Academy in Wasilla, Alaska. She teaches a variety of students who range in ages from seven to fifteen in a single large classroom. Years ago she took a chance and invested in the thirty-four video set in hopes of sowing the seeds of science in the minds of her young students. She reports that out of the minds of babes came weekly viewing that evolved into a set day: *Paul Hewitt Thursday*! Without fail, students anticipate the fun of conceptual physics every Thursday! Nina tells me that the young ones absorb the concepts and continue as little sponges with every passing year. By the time they reach the eight grade, the concepts have

Weekly Paul Hewitt Thursdays

become embedded and students perform well when tested on them. She even hears some very young ones correctly answering Paul's posed questions during the video presentations. The Paul Hewitt Thursday has become a fun

Walter Steiger, me, and Suk Hwang

massaging of little minds and confidence building of older minds. In a word: Yum!

The beauty of Hawaii extends beyond Honolulu— particularly to Hilo on the Big Island. I was invited by UH Hilo physics department head Suk Hwang to teach Physics 100. There I met Walter Steiger, physicist-astronomer professor from UH Manoa, the pioneer of the solar telescope atop Mount Haleakala in Maui, and tsunami expert George Curtis. Both became close personal friends.

While in Hilo I also met glass icon Dale Chihuly in a workshop organized by local community leader Alice Clark: "The Physics of Glass Blowing." It was her way of connecting science and art while also bringing together two people she admired—Dale and me. The workshop was enjoyed by many—it was a blast!

Although I loved visiting Hawaii, being a resident is another story. I felt welcomed by all I met. But I felt uneasy about my intrusion into a culture that in many ways I respected—but was not mine. I was a guest of Hawaii, not one of its residents. For example, I view the Big Island's tallest mountain, Mauna Kea, as one of Nature's gifts to the world—not sacred only to Hawaiians. I cannot agree that being sacred means that its telescopes desecrate the mountain. I see the telescopes as eyes that magnify the magnificence of the universe—for all people. Growing up on the American mainland, I feel greater comfort culture-wise in California, Colorado, Florida, or other mainland states.

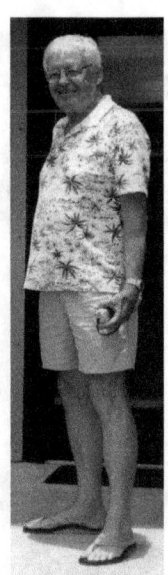

George Curtis

The teaching stints I cite in this chapter were wonders beyond my wildest dreams. Being a visiting lecturer at UC Berkeley, UC Santa Cruz, New Mexico State University, and both universities in Hawaii were fabulous experiences. A wonderful feature of teaching is the chance to travel and to attend physics workshops and conferences—and to meet new friends.

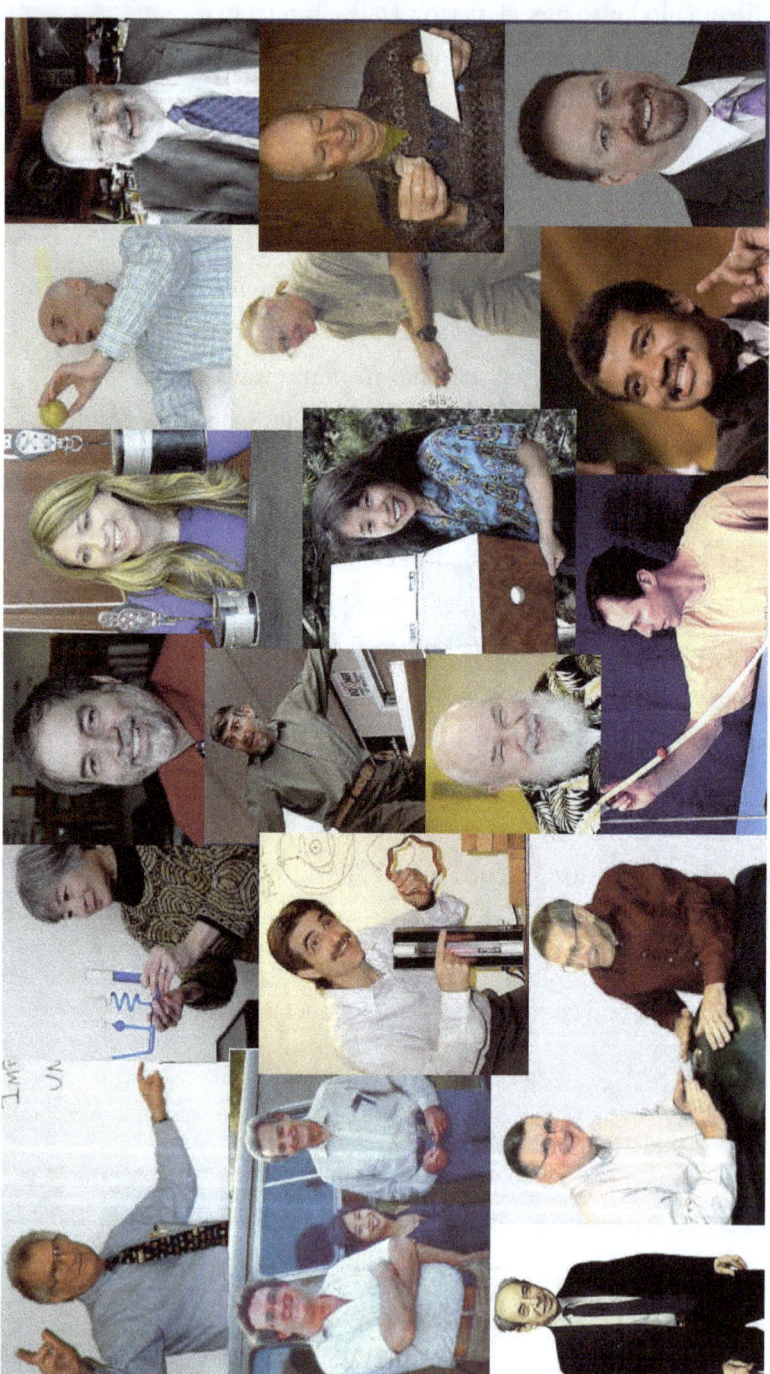

Why are they all happy?

Answer: They are physics teachers.

FIFTEEN

Workshops and Conferences

T he American Association of Physics Teachers (AAPT) is the place to make physics friends. There are two national meetings per year: a winter one in January, and a summer one in July. I usually attended both. Then there are regional ones, for example in northern California and in Florida. I extolled the benefits of teaching conceptually at many of these events. A memorable national summer meeting in 1977 took place in Puerto Rico, when I brought my thirteen-year-old son James. It was my pleasure to introduce him to my physics friends. This was at a time that skateboard popularity was spreading nationwide. James brought his skateboard along and when we visited the huge open-dish Arecibo telescope we pondered his prospects of riding along the surface of the big dish. Of course, this speculation was all in fun.

What goes well in the classroom may not go well at AAPT meetings. A glaring example was the one time I brought my classroom light show to a national meeting. Each semester in my Physics 10 class I treated my classes to a light show with music, all squeezed into my time between other classes in S100. In a booth at the back of the lecture hall I hastily set up a pair of Kodak Carousel slide projectors, while an assistant set up a tape recorder on the lecture table. We played Santana's *Samba Pa Ti*, and Traffic's *Dear Mister Fantasy*, and Country Joe McDonald's *Section 43*, three rock music hits of the time. I'd show a series of slides from the projectors, moving my hands to and fro across the lenses to project swirling images in rhythm with the music. The slides were abstractions of natural surroundings from my collections over the years. Quick setups and projecting a rather rapid series of images, being sure one lens was covered while advancing to the next slide. It was an exercise in coordination. The show lasted about twelve minutes. Students loved it.

But the light show was not appreciated by professors and teachers when I presented it at a Chicago AAPT meeting. Scheduled between professional physics papers, the twelve-minute show seemed to last an hour. I cringed when I realized it was way out of place. What was wonderful in my Physics 10 classroom was an embarrassment with the AAPT audience. A lesson: What works for one group may not work for another group.

Another science teaching organization is the National Science Teaching Association (NSTA). Between AAPT and NSTA, I got to share my ideas with many. I relished showing teachers how to make quick sketches of cartoons to perk up classroom presentation in my cartoon workshops, which became a much appreciated feature. The origin of these goes back an experience at Lowell Tech watching a professor fumbling at the chalkboard while trying to draw a cube. We students had to sit patiently while he kept sketching and erasing until finally a decent cube was drawn. I thought, why doesn't this guy take a few minutes and practice the drawing before class time? Any quick drawing would spice up a lecture. *Quickness* is the key. If not done quickly, students would rather you continue your lesson than watch you carefully display your artistic talent.

In the workshop, usually an hour and sometimes more, I'd show simple step-by-step procedures for quick chalkboard (and later whiteboard) drawings. Attendees would copy my successive steps. A large room with many

A fun time, learning to draw

boards worked best. Some attendees that preferred not to stand at the board could follow along while seated in chairs with pen and paper. I encouraged the board experience because it's common to teach in front of them. What I particularly enjoyed at such workshops was the same thing I enjoyed when creating Figuring Physics pages. That was to elicit, "Hey, that's a good way to say that" in lessons. I think my cartoon workshops were the tastiest of my presentations.

Although my reputation for cartoons was well established, it was satisfying to compare the quality of my drawings with those of the attendees because there were always better drawings than mine! The main difference was my confidence in drawing. It is safe to say that all attendees enjoyed the workshop. Why? Because their joy was discovering they could draw better than they expected of themselves. Isn't this similar to the satisfaction our students experience when they discover they understand a topic they initially thought was beyond their comprehension? It's an uplifting feeling to be good at things!

In 1988 my publisher, Addison Wesley, arranged a visit to their overseas office in Sydney, Australia. From there I attended a National Science Conference in Alice Springs, a remote small town in the northern middle of the country, noted for its Aboriginal culture. I gave my pitch for teaching conceptually along with a well-received cartoon workshop. At the conference I met Alan Pepper, who invited me to his high school in Adelaide where I had

Roger Rassool

Coffee with Alan and Ann Pepper in Adelaide

the chance to speak with many of his students. It was a joy. Then to Alan's home on Hewitt Avenue. Wow! I'd never before encountered a street named Hewitt. About a week later, with a lot of free time, I climbed what was then known as Ayers Rock. In a later visit to Australia I visited Roger Rassool in Melbourne, a physicist that I had previously met in Sweden. Australia is a delightful country, seemingly like America used to be before it became so industrialized.

During the summer of 1992 at UH Manoa I was asked to give a week-long workshop to physics teachers in Chuuk, one of the four states of the Federated States of Micronesia located in the western Pacific Ocean. During World War II, the Japanese used the lagoon in Chuuk, then known as Truk, as their main base of operations in the South Pacific. Japanese engineers constructed a large concrete ramp in the lagoon that dipped from land into the water to accommodate amphibious vehicles. While snorkeling I was amazed to see that the ramp was fully covered with beautiful corals, some round ones nearly a meter in diameter, among others that resembled antlers. I had always wondered how long it took for corals to grow. At this point I had more information about this. Since the ramp was constructed in the 1940s, the age of coral on the ramp couldn't be more than fifty years. I had thought that coral grows

Alan, William, and Fe Davis

more slowly than that. Interesting. My workshop went well, and peeking in was Alan Davis, an American marine scientist. Having much in common, we became friends, a friendship that about a decade later was strengthened when he relocated to Oakland, California. A lover of ocean science, he contributed the tidal chart that graces the pages of my thirteenth edition of *Conceptual Physics*. Alan, his wife Fe, and his son William are cherished friends today.

Near the end of 1992 I was invited to present a keynote address at an international UNESCO conference in Manila. The Philippines is well known for providing nurses worldwide who send their earned funds back to their families. I found in that enchanting country that it also produces great physics teachers. I corresponded with one such teacher, Jules Layugan, who taught at

Jules Layugan Puno

Philippine Science High School in Manila, and became my pen pal for more than a year before we met at the Manila conference. I expected Jules to be a guy, and how surprising to meet not a guy, but a lovely woman. Jules was married with two children! It was December and I had just turned sixty-one. I celebrated my birthday with Jules, and master of ceremonies of the conference Jess Rivas, and Nyrma Lacanilao. Jules and I continued physics chatter online and soon thereafter she immigrated to the United States. Upon becoming a US citizen, she reverted to her maiden name Puno, teaching physics part time and coaching teachers at James A. Garfield High School in East Los Angeles. The Manila conference was the last of my many visits to the Philippines. Photos of Jules grace some of the editions of *Conceptual Physics*.

In mid-June 1996 I was honored to present a two-week workshop at Fermilab, outside Chicago. I was eager to meet its director, Leon Lederman, not so much because of his high standing with physicists worldwide, but because he was an outspoken advocate of a physics-first sequence in American schools. I was greatly impressed that Leon greeted me with a first-rate personal tour of the huge particle physics laboratory. He had recently discovered the long-lived neutral K-meson, for which he later shared the 1988 Nobel Prize in Physics. My workshop with more than fifty physics teachers went very well. I shared with them all the teaching tips I championed. Especially gratifying was receiving a standing ovation on the last day of my presentation. I repeated the workshop with other Chicago teachers the following year.

In July 1996 I joined Richard Olenick, Carl Rotter, and Kelly Woods in their two-week Dallas workshop Comprehensive Conceptual Curriculum for Physics (C3P). This Texas workshop was attended by seventy-two high-school teachers from thirty-eight states. My strong view of teaching is that many teachers teach as they've been taught. Hence, the attendees would be my students, and I'd be their teacher—as if this were their first time in a physics classroom. Although the physics was quite familiar to them, I wished them to appreciate and hopefully adopt my new and different method of teaching. It went well, but I would have preferred that they were younger teachers less set in their ways. A feature of the workshop was assembling vast amounts of teaching material into a single DVD. All bases were covered. But I was less than pleased, remembering my number one enemy of learning—*information overload*. I'd prefer that information be condensed to essential nuggets. I accepted a revisit the following summer. All in all, as my diary entry for July 25, 1997: "Finished the lecture series. Time well spent."

In 1998 some friends and I presented a week-long teaching workshop in Burnaby, B.C., Canada. It was hosted by physics teacher Brian Jackson at Burnaby South High School, close to Vancouver and the home base of my good friend and inspirational physics teacher Peter Hopkinson. It was Peter who suggested I do this summer workshop. My presentation made use of the overhead projector, quite popular at the time. Two years later, a second workshop was held at Vancouver Community College hosted by Peter. My assistants for both were David Vasquez, Howie Brand, and videographer Duane Ackerman. Although the participants were forty or so teachers and professors, repeating the method I developed in Dallas, I viewed the participants as students, which worked out well since most were new to a nonproblem-solving way of teaching. The resulting videos were satisfactory.

Videographer Duane Ackerman, Semi-host Peter Hopkinson, David Vasquez

Conceptual Physics Workshop
in Vancouver, B.C., Canada

Duane nicely selected the highlights, which can be seen in his promotion of the longer workshop on the Internet. I think Duane's promo piece is more than adequate for those wishing to learn some of my teaching tricks. Participants who became friends included Jim Redmond from Oahu, Hawaii, and

Howie Brand and
Mona El Tawil-Nassar

Evelina Chiu   David Riveros Rosas

Julia Bergman, me, Evelina Chiu, Masha Zakheim, and Will Maynez in Mexico

Mona El Tawil-Nassar from Cairo, Egypt. How I wish I had known Mona on my first visit to Cairo earlier in 1972.

I taught summer workshops in Mexico City in 2000. My hosts were Evelina Chiu and David Riveros Rosas who translated and adapted my high-school edition of *Conceptual Physics* into Spanish for the middle-school Mexican Physics Program. Adding to the festivities in Mexico were my friends Will Maynez and CCSF librarian Julia Bergman, who were curators of the famed CCSF Diego Rivera mural. We enjoyed tours of Diego's studio. Later in Mexico City I learned of a great personal shortcoming when I agreed to teach a sample lesson in a TV studio. For me, teaching has to be an interaction with students, even with seasoned teachers taking the role of students as in my workshops in Texas and Vancouver—but not with spectators. The resulting video was a disaster. Although I chose a topic I enjoyed teaching—satellite motion—my presentation was flat. Without the gusto prompted by an audience of students, I was going through the motions—not teaching. I couldn't fake it.

A favorite country to visit was Sweden, particularly in Lund, the southern part of the country. Lil and I attended a 2002 European conference GIREP (International Research Group on Physics Teaching), held at Lund University that began with a trip to the island of Ven and a visit to Tycho Brahe's sixteenth-century castle and observatory. A follow-up feast of Swedish cuisine coupled with professional entertainers produced the conference of conferences, never to

Meeting Per Olof and Anette Zetterberg in Lund

be equaled in our experience. At the conference I gave a talk that must have been very impressive to department chair Per Olof Zetterberg because we were invited the following year and give a talk to prospective students. We stayed at the posh Grand Hotel Lund, all expenses paid. This generous invitation was repeated for two consecutive years. Being a bit wowed that the university picked up the tab for these visits, Professor Zetterberg explained how it was good business. He said that if my talks resulted in only one student deciding to attend Lund University who otherwise may have chosen to study elsewhere, the university would be ahead financially. I was glad to be part of a recruitment effort. Remarkably, in the years 2005 to 2014, Lund had as many Natural Science Physics students as all other Swedish universities combined, which included half of all physics majors. Adding to our good times in Lund were visits to American instructor friends Dean Zollman and Stan Micklavzina. We greatly enjoyed multiple visits to Sweden.

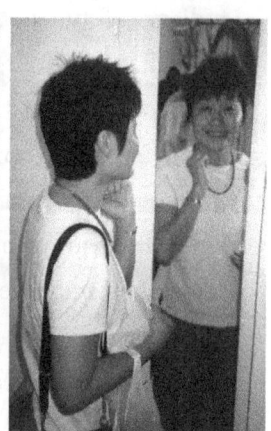

Lil with long hair, standing out, and short hair.
Donating hair to science in Lund University.

Lil accompanied me whenever I visited Sweden. On our first trip she was delighted to get her hair cut by a hairstylist in Lund—nice and short. Later I asked her if she'd grow her hair long for an electrostatic repulsion photo for my next edition. She agreed and the following year when we visited Lund again her hair was nice and long. Per Olof supplied a Van de Graaff generator and took the photo with her electrified hair standing out when she made contact with the charged generator. After we saw that the photo was a good one, she returned to the same hairstylist to have her hair cut short again. When asked why she parted with her lovely long hair, she replied it was only long for science!

Our Swedish welcome extended to the University of Uppsala, north of Lund and Stockholm. This involved an ongoing friendship with Cedric and Anne Linder who previously lived in Capetown, South Africa. Cedric taught from my textbook and emphasized concept understanding in his teaching. He published an article in *The Physics Teacher* about his great success with my book. This occurred when my approach to teaching was seen by many as lowering math standards. I think it was Cedric who first said, "When a learner's first experience in physics is delightful, the rigor of the second course will be better understood and welcomed." It was back-and-forth correspondence with Cedric that led to our friendship. Cedric and Anne later

Anne and Cedric Linder in Uppsala, Sweden

immigrated to Uppsala, Sweden, when Uppsala University wished to implement conceptual teaching in a newly formed division of teacher education. Cedric was the person. All visits to Sweden included the Linders. On one

Derek Muller

such visit was Derek Muller, a friend that my publisher wished to groom as a coauthor for my book should my writing skills wane in coming years. As it happened, Derek reached and influenced a considerable number of people with his website *Veritasium*. Cheers to Derek Muller! I had previously first met him when I guided him around the Exploratorium in San Francisco. It was great to see him again in Sweden.

New Zealand has a reputation of being every traveler's favorite country. Lil and I found that to be true in 2003. We took part in the New Zealand Institute of Physics Conference at Massey University, Palmerston North, where we were graciously hosted by Jennie McKelvie. Then on to Victoria University of Wellington at the southern part of the North Island. At Wellington we met Sir Paul Callaghan, the Director of the MacDiarmid

Chatting with Sir Paul Callaghan

Meeting with Jennie McKelvie and David Housden

Institute. During my speaking engagement there, he realized that Lil was very familiar with my talks and suggested a shopping tour not far away from the university. In addition, he provided round-trip taxi fares. Lil had a ball. Of all the conferences Lil attended, she rated Paul Callaghan as the top host. We couldn't have been better treated. Our friendship sadly ended when Paul Callaghan passed away in 2012. But another friend emerged, high-school physics teacher David Housden, who hosted much of our tour of New Zealand. David's expertise is authoring exams for the best and the brightest high school physics students in New Zealand. We've since exchanged teaching ideas via email over many years. Lil and I decided to save the South Island of New Zealand to ensure a future trip—which hasn't yet happened.

After I gave a plenary talk called "Developing Conceptual Physics" at an AAPT winter meeting in Baltimore, Maryland, in 2008, an attendee in the audience asked who was my most outstanding student. My answer was a quick one: Tenny Lim, my top CCSF student who went on to earn a mechanical engineering degree at Cal Poly in San Luis Obispo. I told Tenny's story. During an interview for Jet Propulsion Lab (JPL) employment, her interviewers were impressed that her college transcripts cited art courses each term along with engineering courses. When asked why, she replied that art was her passion. That was just what JPL was looking for—someone to learn

Tenny and hubby
Mark Clark

Tenny with a model of the Curiosity rover

the image-making promise of computer graphics in addition to having technical savvy. To celebrate my upcoming eightieth birthday, Lil and I, with Tenny and her husband Mark, watched the launch of the spacecraft that carried the Curiosity rover to Mars. It was a special occasion for Tenny—for she was the lead designer of the technology by which Curiosity was safely lowered onto the Martian surface. Since then, in my classes I'd stress the lesson from Tenny's experience: Excel at more than one thing.

I've always believed that almost all teachers begin their career wanting to impart the information they grappled with in their student years and to share it with an eager class of young students. If they connect well at teaching, then they will receive student appreciation. For some of those who don't relate well, the hoped-for interaction with students morphs into being adversarial. Their mood changes to "if they don't appreciate me, they'll at least learn to respect me." Or in the extreme case espoused by a very unpopular teacher at CCSF, "I don't like this any more than you do, so let's roll up our sleeves and look at the problems at the end of chapter so and so." Wow! This sorry scenario doesn't occur for teachers who know and explain their subject well and have the final touch of caring about their students. I like to think that we

Hanging out with Ton Ellermeijer and Ed van den Berg

all know that. The lesson? Caring for students is essential, but not enough. We must couple that with our utmost skills to instill student learning.

At an annual Netherlands Teacher Conference, I met a fellow who introduced himself as Ed van den Berg. Wow, I replied, it's been years since you taught me to play guitar in my Army years. My Army buddy Ed Vandenburg of many years ago had the same name. So how nice for me that there have been two Ed Vandenburgs (okay, different spellings) in my life! Ed and I hit it off and have been close friends since. His wife, Daday, is also a physics teacher and close friend. Together they taught physics in teacher education programs in the Philippines and Indonesia. In their projects they encouraged the very top students to become physics teachers through special Science Teacher Education scholarships. This model worked well in the Philippines, though unfortunately not as well in richer countries. I was not surprised to hear them say that teaching in poorer countries was more rewarding to them. It was gratifying to learn that their model of teaching centered on my textbook. The van den Bergs invited Lil and I in 2009 and again in 2016 as guests in their home in colorful Heiloo, not far from the tulip blooming farms.

Other Dutch friends were Ton Ellermeijer and his wife Ewa Kedzierska, of the AMSTEL Institute at the University of Amsterdam. Ton and Ewa were more than top-notch hosts in that wonderful city of canals. We were pleased when they accepted a visit to our home in Florida in 2004. During the

Christmas holidays of 2009 they treated us to a memorable tour of Holland's famed windmills and towns outside of Amsterdam. Again, a valued sidelight of teaching is making great friends.

At a 2009 AAPT conference I met a local high-school physics teacher Scotty Graham, who brought a few of his students with him, which was quite unusual. He and his students had built an organ as part of their study of musical sounds. A year later he traveled to Manila in the Philippines to see and hear the world's only surviving bamboo organ. He found much more than organ music to appreciate. He discovered an uncommonly friendly culture and wished to retire there. When

Scotty Graham and family

returning to teaching in Texas he suffered a stroke. After a slow recovery he made his way back to the Philippines in 2014. He never returned to the United States. He married a local young woman and had two children. Scotty has graciously contributed material for my writings. More than that, he is now my valued pen pal in the Philippines.

In 2011 we attended a GIREP teaching conference in Finland. I gave my usual talk along with a cartoon workshop. In Jyvaskyla we met Z. Tugba Kahyaoglu, a physics teacher from Turkey, who has become a cherished

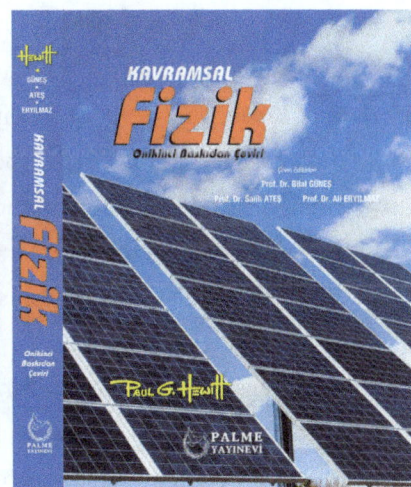

Z. Tugba Kahyaoglu   Helping students

friend. We have since visited her in Istanbul. When my twelfth edition was translated into Turkish by Bilal Gunes, Tugba translated my *Next-Time Questions* into Turkish, available on the web to all.

Our most remarkable experience in Finland was a conversation we had when having lunch in the university cafeteria and chatting with a young high-school teacher. He told us that Finnish students rate highest in international testing without stress because education is not centered on exams. Nor is there homework. Class sessions are among the shortest in the world and summer vacations are longer than elsewhere. We were impressed when the teacher told us that the salaries of high-school teachers in Finland were very nearly the same as that of professors at universities, and furthermore, teachers are among the most highly respected members of their communities. He further suggested that the road to becoming a teacher in Finland was more difficult than the road to becoming a lawyer in America. Wow!

Only Finland's best and brightest have a chance to enter the teaching profession. First, they must be above the top 80 percent of their class in secondary school to qualify for acceptance to the School of Education. Whoops, said my wife Lil and I, in our country *all* students, struggling or not, are welcomed to departments of education for teacher training. In fact, if students can't succeed in a hoped-for major, they are always welcome to enter the field of education. In contrast, I remember while in graduate school at Utah State University unfortunate physics students were called into the chairperson's office and given a "no-thank-you" handshake to exit the program. I also learned this never occurs in the education departments. Every student succeeds in becoming a classroom teacher. That's true in the US—but not in Finland. So why do Finnish students, along with international students who are welcome to study in Finland, attain the highest scores on international exams? Why do they perform so well? The primary answer is simple: They have better teachers.

So we get what we pay for. Americans in general today aren't willing to pay the wages of Finnish educators. Another reason is that our best and brightest students do not want to endure the day-to-day routine that teaching requires: creating lesson plans, grading exams and homework galore, problematic student behavior, not offending unruly students, and on top of that, having a low standing in the community with inadequate pay. I like to think that in time our schools will morph to a Finland-type model. We can begin by attracting bright teaching candidates with pay scales that match the earnings of the best and brightest in industry. Furthermore, the teaching candidates, as in Finland, must be compassionate toward students. Students would then have teachers who not only know who clearly explain a subject, and genuinely care about them—who would treat them as family. Abiding by agreed-on goals, teachers would be given free rein in the classroom. Highly intelligent skillful teachers know, or soon learn, how to educate their students. My hope, rather than my expectation, is that this will happen—but not tomorrow. Rome wasn't built in a day.

On a visit to Norway in 2012 it was pleasant to meet Ole Anton Haugland who teaches physics at the University of Tromsø. He arranged a delightful four-day boat cruise from Bergen to Tromsø, north of the Arctic Circle and home to the world's northernmost university. After we arrived and booked into a Tromsø hotel, Ole and his wife Randi drove us to their house for some yum Norwegian home cooking. We had hoped to see the aurora borealis, but the skies remained overcast during our time there. If we had extended our stay, we would have seen the aurora the next night. (Lil and I later witnessed what

Ole Anton Haugland
and Aage Mellem

Carl Angell

we missed with the Hauglands on my December birthday in 2018 in Iceland.) Ole Anton has contributed photos of the aurora for my books.

Farther south in Norway we met esteemed teacher and physics textbook author Carl Angell at the University of Oslo, and we discussed the elements of good teaching. I'll always remember Carl adding to the list, "You've got to love them." So true, which was the essence of Ken Ford's 2006 Oersted Medal address. Ken's title for his response to the prestigious award was "Love Them to Death." Loving your students as family is an essential feature of successful teaching. Some teachers find this easy, but regrettably, others see it as an uphill climb.

Back to America. Just as too many cooks spoil a stew, so it is with teaching. I recently learned of a frustrating situation from Jan, a seasoned first-grade teacher with great success in teaching math to first graders. It seems that an "education specialist" was hired in her district office. The specialist has a degree in education and earns a living developing new methods of teaching math. And to justify the new ways, tests are created to assess the new methods, which assures that the specialists remain on the payroll. This means that Jan must suspend her way of teaching and implement the new and hopefully improved method in her classroom. Jan's response, "If it's not broke, don't fix it," falls on deaf ears. More specialists with educational degrees are employed to sort the data generated. So money allocated to "reinventing the wheel" in Jan's school does not go into providing classroom teaching assistants, which Jan and her teaching colleagues would love, but funds are vaguely directed to "education overall." What a contrast with Finland, where the well educated Finnish teachers have the smarts and classroom savvy to freely decide how best to teach their subjects.

Another frustration occurs in American education—continual revisions of state and national standards—too many cooks spoiling the stew. I've been told by reputable colleagues that the choice of topics need to be in a curriculum are often crafted by people who are the least qualified to decide which topics are essential. Nevertheless, these policy makers produce the new standards. An upside is that when implemented, school administrators can check to see that the books they adopt pass muster with the new requirements. This would be helpful in keeping books up to date. But not helpful, for example, when a teacher is denied using a book by Einstein for teaching relativity because Einstein's treatment of relativity fails to match the new standards. Or closer to home, my *Conceptual Physics* textbook and program being barred because there is no coverage of a certain type of transistor in its electricity chapters. I have to wonder how similar this widespread practice relates to what Jan experiences in her first-grade classroom. This bothers me.

Another problem that I see currently is influential educators placing undue emphasis on science process skills. These certainly have their place in science curricula, but not if they crowd out essential content. Recently I learned that a class spends *weeks* developing models of the transistor. Uh-oh—classroom time is finite, which indicates model-making activities took the place of essential electricity content. Let's not say good-bye to a content-based physics course! This would mean basics such as Newton's laws becomes valued little more than material for classroom exams or labs, and not much beyond that. Students weak in content should be a no-no. Just as my beef with kinematics is that it swallows up too much time, likewise with teaching time-consuming process skills. The same is true of chemistry and biology. Today's gooey impediments didn't exist in my day. I was lucky to teach in what I considered to be the golden age of teaching, which ran parallel to the exponential rise of industries within the science, engineering, and computer technology.

Well into retirement in 2014, Lil and I created 149 illustrated physics lessons for students and teachers. We were motivated by the popular Khan Academy, where Sal Khan tutored his far-away cousin online and then gained an international following. Our lessons were less detailed, and aimed at average-ability learners who valued slow-paced key-concept explanations in an elementary physics course. All tutorials were accessed by quick response (QR) codes in the margins of the twelfth edition of *Conceptual Physics*. None are paired with pesky advertisements and remain available on the web free to all. Many years ago, my thirteen-year old Exploratorium student Toby Jacobson was my first webmaster and created ConceptualPhysics.com. Years later physics teacher Lynda Williams kept the site active followed by teen nephew Christopher Lee. Most recently, nephew John Suchocki keeps it humming.

To end this chapter on a high note, I'll say a bit about the personal adoration that teachers can receive. Mine from meetings with student groups over the years were exceptional. Back in the 2000s I was invited to teach a class at a high school in Connecticut where I was an unknown. I had some difficulties getting the students to pay attention to me. They knew nothing of my books or my classroom videos. That same afternoon I visited a nearby all-girls high school and was overwhelmed by students joyfully yelling while running up and down the school stairs upon hearing of my arrival. They knew me from my videos. After my talk, girls lined up for autographs on notebooks, binders, and even the underside of one girl's skirt. To them I was a movie star. More recently, and into my retirement, I met with students at school physics clubs, some of whom sported caricatures of me on their T-shirts. In one group I was referred to as Captain Good Energy. At another gathering of fans, a girl announced that I ought to be cloned, not only to provide more teachers, but to be spread around for girls looking for good boyfriends. Wow! It would have been unimaginable for such to take place during my earlier insecure skinny years. Instead, this occurred in my seventies, maybe close to the tipping point of becoming just another old man talking. My life has truly been awesome!

Captain Good Energy

Physics club enthusiasts

SIXTEEN

Exploring the World

From the beginning of my teaching career in 1964, I taught Physics 10 nearly every summer for seven years. Simply stated, I just loved teaching. Just as a stage performer loves connecting with an audience, so it was with me—but very much more. Connection with my students was fulfillment at its highest—me teaching, and them learning—particularly because many of them thought that physics was beyond their abilities. My joy was eliciting the student exhilaration that comes with actually grasping the physics concepts, day after day. But then my summer sessions came to an end. They were replaced by world travel.

At age forty I had yet to experience foreign travel. I realized this in a coffee shop in downtown San Francisco while overhearing people chatting about their adventures abroad. Whoa, I thought, this is a gap I should fill. I should travel. What country would be a good start? I thought of the UK, particularly London, but decided it would be a version of the place I grew up, Boston. I wanted a different experience. Aha, graduate-school friend Ellen Drake had located to Ethiopia and quite likely was open to a visit.

In the summer of 1972 I was off to Ethiopia. Before leaving, a student in my class urged me to schedule a stopover in his hometown of Cairo where he'd meet me at the airport and give me a short Egyptian tour. I took his advice and arranged a stopover in Cairo for a couple of days. Alas, my student was a no-show at the airport! I carelessly had no plan B and felt quite helpless. I resisted local boys wanting to help me with my luggage, and I hitched a ride with fellow airline passengers to the big city. Car horns were blaring everywhere, a custom very new to me. In downtown Cairo, my airline companions left me to fend for myself. A local shady-type pest approached me who persistently insisted on being my guide. I managed to lose him for a bit and checked into a hotel. From my hotel room I stepped on the balcony and gazed at the activity below—throngs of people and camels with a strong fragrance that defined this new foreign experience. I was excited, and also lonely.

My host Ellen Drake

All was so different. Then I looked up at the Moon above. Ah, the familiar Moon, the same one I'd known all my life—a companion. My lunar connection was comforting. Most of the next day was spent at the famed Grand Egyptian Museum, where I viewed the King Tut tomb, the same one that toured America some years later. No crowds. All delightful. The next day a plane ride brought me to Addis Ababa, the capital city of Ethiopia. I was eager to meet Ellen.

Seeing the poverty in the streets of Addis Ababa was shocking—very shocking. Walking along the streets I was surrounded by curious children, which I found delightful. Other travelers told me I'd soon find swarms of kids a nuisance. That didn't happen for me. When I asked why a man on the bridge was crying, the children simply said he was hungry. Walking further, I came upon a construction site and watched workmen constructing a three-or-four-story cement building. They were skinnier than me. They also seemed unhappy. I wondered about them, whether their wages were enough to provide a fair enough life for themselves and their families.

Several days later I learned more. Ellen was a friend of Stanford graduate Paul Selassie, the crown prince of Ethiopia and the favored grandson of emperor Haile Selassie who still ruled the country. I was lucky to be invited to join Paul and Ellen in an automobile tour of the city. When passing the construction site that I was previously curious about, I pointed to the workers and asked about the wages they earned. I expected Paul to say something to the effect that when it was his turn to govern, he would do his utmost to ease the poverty of workers and lift the country economically. He did not say that! What he said astonished me and left me speechless. He bluntly replied that because Ethiopia wasn't burdened by the union problems of the United States, buildings could be built for a pittance. Most troubling was his tone, one of contempt for workers he viewed as dirty and despicable, very unlike the Ethiopians that Paul Selassie hobnobbed with—educated, clean, and articulate in conversations. I was awestruck—but enlightened. This explained politics! Such leaders had little desire and no plans to raise the living standards of a workforce under their control. Their focus instead was to ensure the continuance of wellbeing for the few that benefited by toils of the filthy masses. This sordid view of politics was educational. It has never left me.

Ellen and I took a small airplane ride to see one of Ethiopia's big attractions, the centuries-old churches carved out of solid rock in the town of Lalibela. Flying over villages in the countryside we saw many round dwellings and some rectangular ones. We could estimate the age of a village by its roofs. The round ones were cones of straw and the rectangular ones were newer dwellings constructed with sheets of corrugated steel. In Lalibela most all were round. I was fascinated with the unique churches, but also talking with children who gathered around us. When an interpreter told them that I was a teacher they jumped with joy. They begged to know if I could take them with me to attend an American school. I told them that wasn't possible

Church of St. George
in Lalibela

Lalibela cross

Homes in Lalibela

for many reasons, but I'd tell my American students how much the Lalibela children would love to join them. And when I returned to CCSF, I did just that. I told my students that their seats in the classroom would be precious to the youngsters I met in Lalibela. This small story seemed to make a good impression on my CCSF students. They did know *how* lucky they were.

Later I flew south to Kenya. While in Nairobi, and a passion for snorkeling, I was interested in visiting a coastal city in Africa. A delightfully long train ride later from Nairobi I was in that city by the sea, Mombasa. Where to stay? Being a novice at travel I played it by ear and found a decent downtown hotel for eight dollars per day. But that night I had problems with mosquitoes. Hey, this was malaria country and I had to find safer quarters. From the train ride I had seen the large billboards that advertised the upscale Nyali Beach Hotel in Mombasa, but that option didn't fit my budget. Nevertheless, after my encounter with mosquitoes I made a phone call to that hotel and was lucky that a room was available. I didn't ask for the price; staying healthy

came first. A taxi took me there and I was aghast when checking in to learn the cost was nine dollars per day—with full board. That meant nine dollars for breakfast, lunch, and dinner as well. Furthermore, at night attendants tucked me into bedding with a mosquito net. Morning was met with a knock on the door and a hotel employee with coffee or tea. I stayed for a glorious week.

The Nyali Beach Hotel experience left me with a saying: "The best costs more . . . but often only a little bit more." Then I thought of the many cross-country trips years before, being sure to avoid the expensive looking motels to save money. How much money? I had to discover common sense in Africa? Give me a break. For perhaps an extra dollar or less in my student days I could have treated my family to much more comfortable trips across the country. Another message: We learn too late the simple lessons in life.

While in Mombasa I saw Tiana, an American young lady that I'd noticed on the train ride from Nairobi. We spent a couple of hours taking in the sights together before she continued her travels south. She was black and sported an Afro hairdo that was common with my black students in San Francisco. Interestingly, I saw no blacks in Africa with Afro hairstyles. While walking along a Mombasa street where women were selling trinkets, I was astonished to see some of the women spitting at her. Tiana told me that she was used to it—that African blacks look down upon American blacks. I found this to be true in other instances during my time in Africa. This bothered me.

When I later returned to Nairobi, a great stroke of good luck occurred when I met Jack Hopcraft, a naturalist and friend of Ellen's. I met him in his office when he was having a troublesome phone conversation. He had just learned that he needed a driver for an expedition that was in progress. After the call, when I introduced myself, he asked if I had a driver's license. Jack needed a driver for a group of Americans already on a photo tour that I may have heard of—the Sierra Club. Would I accept? No wages, but no cost to join the group who were headed for the Serengeti to photograph wildlife. He was delighted with my quick answer, "Yes, when do I start?" Tomorrow morning was the answer. I checked out of my hotel and he drove me to the group the following morning, suggesting that to complement driving I give fireside lectures on the stars and such.

This was almost too amazing to believe. Jack delivered and introduced me to the group of mostly Americans for the second week of their two-week tour. Each night we drove to a site that was prepared by Jack's employees, who pitched tents and prepared tasty food and wine. One of the tourists who happened to love driving took my place at the wheel. On some evenings I gave lectures very similar to those I gave at CCSF. We zipped our tents securely as curious wildlife investigated these human visitors. To conclude this marvelous expedition we hiked to the snow line of Mount Kilimanjaro. This was one of the most adventurous weeks of my life.

My travels continued in 1973 as part of a sabbatical leave from CCSF. Whereas it was commonplace that a sabbatical leave meant continued study or gaining research experience, my stated reason was to visit some of the countries my students were from. It was accepted. My first stop was Japan, a brief visit where I was enormously impressed with how wonderfully Japanese people treated guests. When I asked directions, the person I asked would take me there! When a bank couldn't cash a traveler's check, the bank teller escorted me to his car and drove me to a bank that would cash the check.

What a caring culture! I decided that I'd have to return later, which I did, several times.

Then to Hong Kong where friends of CCSF students awaited me. I stayed in hotels in Kowloon and on the main island of Hong Kong. I was impressed with the great amount of building going on, particularly the construction of underground rail transportation along Nathan Street in Kowloon. Bamboo, not steel, was used for the scaffolds in erecting buildings. Whereas most travelers spend two or three days in Hong Kong, I stayed for three weeks. During that time I visited the physics departments of Hong Kong University and delighted in chatting with fellow teachers. I did the same in nearby smaller schools.

I was not aware of the small living quarters of Hong Kong residents. I bought lots of things in Hong Kong and wished to store them while I spent most of the summer traveling to Taiwan and the Philippines. I asked a new friend if she could store my small duffel bag during my couple of months of travel. At first she seemed to resist my request, but then graciously said she'd care for my bag while I was away. When I returned, she let me into her home and I was shocked to see my duffel bag on top of the foot of her bed. On the nights during my absence she had difficulty extending her legs full length in her bed! What an unfortunate way to learn that living spaces for average Hong Kong residents were exceedingly small!

I spent a lovely week in Taiwan. It nicely began with a visit to the National Taiwan University in Taipei, where I was warmly greeted by both faculty and students. It was the first time I saw fish heads in a student cafeteria. I discovered the dim sum luncheons in the city, and when I later told my San Francisco student friends about them, I learned that dim sum is served at local Chinese restaurants, but in mornings and midday—not in evenings. So I learned in Taiwan what I should have known in San Francisco, which amused my CCSF students. In the capital city of Taiwan, Taipei, and other communities I saw poverty—not as great as I'd witnessed in Ethiopia three years earlier, but poverty nonetheless. This was when Taiwan was manufacturing and exporting clothing to the world, mainly underclothing. It succeeded, for when I returned ten years later, I was astonished to see a much more advanced country. Developing countries in time can become prosperous. Hooray to that! From beautiful Taiwan I traveled to the Philippines.

My arrival in the Philippines began wonderfully. I took a taxi from the Manila Airport to downtown Manila and didn't have small currency for the taxi fare. In our conversation the driver knew I was an American teacher, and said "Welcome to the Philippines, the fare is on me." Wow! What a welcome! The next day I visited the University of the Philippines. I was warmly greeted and agreed to giving a faculty symposium the following week. I later did the same at Ateneo de Manila University. My book was widely used in the Philippines and since I held its copyright, I boosted its use by granting my hosts permission to photocopy, print, and distribute it to schools for the cost of printing. My royalties, in effect, went to students.

My favorite city in the Philippines was high-altitude Baguio, called the city of pines. It was cool there, especially in summer months when Manila was uncomfortably hot. Baguio is about 200 kilometers (124 miles) from Manila. I enjoyed exchanging teaching ideas with the faculty at the University of the Philippines Baguio, as well as high schools in nearby provinces. I returned to the Philippines in following summers.

Onward to Singapore, which unknown to me had strict rules for travelers. I didn't realize those rules until new friends I met at a local high school expressed amazement that I passed customs with my long hair and beard. I guess I arrived at a not-too-strict time. My few days in Singapore were wonderful. It was the second time, after Ethiopia, that I ate food with my hands without utensils. More common were the open-air restaurants with great varieties of Asian foods the city is famous for. I even stopped by the famous Raffles Hotel for their signature drink—the Singapore Sling. An interesting travel discovery in Singapore was the smooth traffic flow, very much helped by the policy of "boxing" busy intersections. The boxed junctions were conspicuous marked with bright yellow crisscross patterns. The rule was stern: if for any reason a vehicle blocks the flow of traffic while at rest in the intersection, a fine citation is served to the driver. Although traffic gridlock was a headache in San Francisco and other big cities, it was rare in Singapore. When back in San Francisco I attended a Saturday "court" for traffic infractions, something I experienced now and then. In the presence of about thirty or so other offenders I told of the Singapore gridlock solution. I suggested it apply here and asked for a show of hands of support. I was a bit stunned when the loud response was a unanimous NO. The general objection was, "What if it were *me* in the box!" Uh-oh, another example of *me* trumping *we*!

From Singapore I progressed on to Nepal to do some mountain hiking. On that flight our plane was diverted to an overnight stop in Calcutta, India. It was a hot night as I and other passengers were bussed from the airport to a hotel. What I saw exceeded the poverty I witnessed in Ethiopia. On both sides of the roadway to the hotel were hundreds of people sleeping next to one another, all aligned perpendicular to the road. Hundreds more not yet sleeping gathered near our hotel, a fancy one built in colonial times. Hotel personnel cleared a path so we could check in. My room on the second floor was huge, with the largest bathtub I'd ever seen. Looking at the throngs of people from the window I realized what distinguished me from them was mainly the American passport in my back pocket. How many in that crowd would be top students in my physics classes? How many would be great teachers? Without the passport and suitable funds, those questions couldn't be answered. I was moved by what I witnessed and slept poorly. The next morning a sumptuous breakfast was served.

My tour of Kathmandu in Nepal was more adventurous than I had imagined. In the middle of this intriguing city is a huge temple, the one that everybody has seen in photographs, a huge dome with a large eye on each side of the rectangular top. I mingled with locals and when I asked about entering the temple, I was amazed

Stupas at Swayambhunath in Kathmandu

to learn that people do not worship inside. The place of worship is the world on the outside—the world of nature. I was very impressed. I also found it interesting to learn that many of the tourists were women traveling alone. Perhaps between jobs in their native lands, they selected Nepal as part of their travel itineraries. When I asked how long they would stay, the usual answer was, as long as their saved funds lasted. Those were the days of open-ended airline tickets.

One of the features of Kathmandu was its pie shops. Fellow travelers raved about them. I sought one out and was surprised to learn the pies weren't some Nepalese delicacy, but were the pies one gets in the United States or Europe. Lemon meringue was the favorite. A bigger feature of Kathmandu were the dens that specialized in marijuana and hashish. Jars of marijuana buds were arranged like jars of candies back in the States—all very affordable. Hashish was in the form of balls, each graded for quality. In the States a piece of hash about the size of a thumb would last many months by shavings to blend with hand-rolled joints. This amount of hash accidentally dropped on the floor of a pot den in Kathmandu would be ignored, which would have been shocking back home. It was pot heaven, safely away from the reach of the US government. The camaraderie with others was as yum as the pot. And for the record, just as I avoided intoxication by alcohol, I avoided the drugs that were addictive. Pot and hash were not. Common sense and a respect for good health accounts for my lifelong choice to avoid problematic substances.

I spent one night at Kathmandu's classiest hotel, Soaltee, for fifteen dollars a night. Clean towels and all, but socially boring. I and other travelers much preferred the grungy very-low-priced local ones. Prices ranged from fifty cents for one where the bed was simply a batch of hay, to a few bucks for ones with clean beds. Interesting acquaintances were a part of these more entertaining places. Hotel hopping was the mode. At one of these cheap hotels I first met Texans Rose and Tim Gardner, who remained friends for several years. They spoke highly of an American traveler, Lori Parisi, who I missed. Lori and I were traveling in opposite directions. Later she married Milo Patterson. Again, a primo benefit to travel is making new friends.

I took a short plane ride to neighboring Pokhara, Nepal's second largest city—actually, more like a town. I was lucky to find a vacancy at the upscale Fishtail Lodge. It's located near the Jameson Trail, the best high path for viewing the picturesque Dhaulagiri. Hiking up the trail was a pleasure. Due to the hot weather,

Fishtail Lodge in Pokhara

my pajamas served as my pants. Tea shops were frequent along the trail. It would have been nicer if cold drinks were available, but the tea was fine for combating dehydration. The shops nicely served as locations to meet fellow hikers coming down the trail. I relished hearing of their adventures. The locals were excited about the new small orange plastic bags that could hold not only small items, but liquids that wouldn't leak—a big improvement over woven grasses or paper. Plastic had arrived in this part of the world. What wasn't appreciated, however, was the continuing overcast skies. At day's end and time to sleep, locals provided sleeping space in their homes for a small fee. When I asked the owner of one such place how it felt to be living in peaceful Shangri-La, he laughed at me, and spoke of the ongoing squabbles with his neighbors. This, he said, was certainly not the most peaceful part of the world! Legend and reality sometimes catch up with each other.

I continued my upward trek early the next morning when the Sun broke through part of the clouds. With my binoculars I couldn't view Dhaulagiri, but I could see the city below. There were two locations in Pokhara that served cold beer—a small shop along the main street, and the Fishtail Lodge. As if I got a message from some great beyond, and wondering how long it would take going down to where the beer was, I made a decision. My upward trek would stop where I was. That same day I descended the trail and was rewarded in the evening with an ice-cold beer. To see the beautiful Dhaulagiri free of cloud cover, I bought local postcards.

The timing for the Nepal visit was fortunate, for the country had just welcomed vehicle travel the previous year. The narrow streets were not suitable for the traffic at hand, let alone that which was sure to follow. Feeling that my marvelous experience in Kathmandu was one that wouldn't have the same value with the inflow of tourists and vehicles in successive years, I never returned. I didn't open myself to that perceived disappointment. On the wall of my home today, however, is a beautiful rug that I purchased at a Tibetan refugee camp on the outskirts of Kathmandu. It's a daily reminder of that extraordinary visit.

My next stop after Nepal was India. July is hot in India, except in the beautiful mountains of Darjeeling and Kashmir. Because I heeded the advice of a teacher friend's sister who was new to being a travel agent, these two high-altitude cooler locations were not on my air ticket. Instead, my time in India was in Varanasi on the Ganges River, where I saw the many rituals it's famous for. I also visited Agra, where I saw the Taj Mahal at the time of a full moon. Both places were wonderfully exotic. The only university I visited in India was without prior notice. I eavesdropped on a physics class and was upset to see the professor reading aloud from a script as he paced to and fro in the lecture hall. The students took notes. To me it was more a visit to a church than a university. I now think that I chose a bad sample—a different university may have been more impressive. Then to Bombay where I treated myself for a few days at the Taj Hotel. It was fifteen dollars a day, but hey, we only live once! And then the monsoon struck. Rain galore. That's when I returned to Nairobi.

My agenda back in Africa was to witness the acclaimed "longest solar eclipse in a century," at Laisamis, a small town in Kenya. Ellen and I rented a vehicle from Jack Hopcraft. Seeing the eclipse was awesome. Also awesome was the group of eclipse seekers that Ellen and I met, people from different parts of the world—a nice benefit of traveling. We

1973 eclipse at Laisamis

pitched tents awaiting the event. I was surprised to learn that when the Moon began blocking the Sun as it moved across the sky there was no need of a flashlight to adjust my camera for good photos. As the surroundings became darker while progression occurred, our eyes made corrective adjustments. During totality I could see my surroundings quite well due to the brightness of the corona. Most fascinating were the thin quivering slivers of strings that emanated from the solar surface. Those quivering strings of light wouldn't show in a camera, for the wiggles would be numerous and blend in any quick camera setting. These vibrating quivers of light could only be seen with the best optics available anywhere on Earth at the time—the eye!

What I didn't know in the summer of 1973, was to look down on the ground for the "sunballs" rather than to look up at the eclipse. These, to me, are more intriguing than what's happening above! When sunlight passes

Pinhole images of the Sun

through the openings between leaves in a tree, those openings behave as "pinholes" and cast pinhole images of the Sun on the ground beneath. During an eclipse, the round images of the Sun slowly transform into crescents. We know that teachers don't know of "everything" when teaching. At that time I was unaware of the circular images cast by the Sun through openings in tree leaves. With regard to "not knowing," I'm reminded of Richard Feynman who puzzled people when claimed he didn't know anything. Feynman meant that what he *did* know is closer to nothing than what it was that he *could* know. He knew enough to realize he had a small handle on an enormous universe still full of mysteries. I loved that guy. Since my many years of ignorance about sunballs, they've become one of my favorite physics lessons.

We returned to the Norfolk Hotel in Nairobi where we joined some of Ellen's friends. On my own I ventured to the Treetops Hotel, famous for its view of wildlife galore on a water hole. All that I experienced in Africa was wonderful. Ellen returned to Ethiopia and I continued my travels. My stay there occurred at an exciting time for its guests. A rare bonobo came into view in the middle of the night.

Up the coast from Mombasa is the island of Lamu, a favorite destination for travelers due to its beauty, and famous for its elegantly carved doors that go back to earlier Arab inhabitants. I planned for a short visit, which became several days due to unusual choppy waters that cut off its boating operations. I and other visitors were stranded for a bit. One evening a group of us with the aid of some sort of drug, maybe Quaaludes, sat on the huge branches of an enormous mangrove tree. Conversations were stimulating, mainly due to a fellow who was into computers. I vaguely recall he was from the San Francisco Bay Area but I never got his name. He talked of making computers small enough for people to have personal ones in their homes, where they could access enormous amounts of information. I joined the conversation by saying I didn't see their value because public libraries already store that information. I failed to realize that easy access to valued information would make the big difference he talked of. How longsighted he was, and how shortsighted of me. Since then I've pondered that I might have been at that tree with computer pioneers who likely worked with Steve Jobs. The revolution in personal computers was in the future—and as I later learned, the near future!

Back to Nairobi where I flew to Athens in Greece. I wanted to swim in the Mediterranean. At a sandy beach on the seashore, I asked local people where I could put my valuables. I was told to simply place them anywhere on the sand. Nothing would be stolen. That may have been because at the time Greece was under a totalitarian government. Thievery was not tolerated. On another occasion I mistakenly left my Pentax camera in a restaurant. Hours later when I discovered my loss, I returned to the restaurant where it was given to me. The most impressive part of my world tour was the island of Corfu. What I most loved about it was the winding narrow streets and absence of street traffic in wide stairways that had been there for centu-

ries. Small, charming shops and places to eat were everywhere. I fell in love with Corfu. Then a day in Paris and a long flight back to the good ole USA. I can't see how my sabbatical could have been better spent.

Travels in the summer of 1974 were with my three kids on

Summer travel with James, Leslie, and Paul

Repeating Galileo's famed experiment

With my kids in Pisa, Italy

a tour of Europe. Before embarking we purchased special tourist Eurail open tickets that enabled much freedom and convenience of travel. No standing in railroad station lines for ticketing. Yum! After a transatlantic smooth flight we began in Paris and traveled by train to many places, including a special visit to Pisa in Italy. At that time, no restrictions were in place to keep visitors from doing "Galileo drops" from the top of the Leaning Tower. I stood where the master may have once stood and dropped a heavy and a light piece of wood. My kids below attested they landed at about the same time. Our summer concluded with a delightful week-long cruise of the Mediterranean. When aboard the Stella Maris we were seated among other tourists, one who introduced himself as Doctor So and So. I said how nice, because one never knows when a skinned knee or whatever would occur, to which the guest briskly stated that he was a Doctor of Chemistry. Uh-oh-an academic, like me—but arrogant. That's when I wished I had a doctorate and could reply that on this vacation I left mine back at City College. Feynman, Oppenheimer, and Ford never called themselves doctor! But I kept quiet so as not to be a "sour grapes" guy. We enjoyed the cruise, especially when James won a dancing contest for his hilarious moves to songs of the Bee Gees. The movie *Saturday Night Fever* was popular then.

Back to my home base in San Francisco. As wonderful as travel is, it's always nice to be home. I enjoyed telling my Filipino students how much I loved their native country. Some gave me addresses of people to look up on my next visit, which occurred in the summer of 1976. So it was back to my favorite city in the Philippines where I spent a wonderful week at one of Baguio's finest hotels, the Crismar Mansion Hotel. I always took my work with me, and during that week I wrote the first iteration of the chapter on general relativity. The hotel owner was Colonel Marcos, a World War II hero, not to be confused with President Ferdinand Marcos. He was impressed with my book, especially knowing that I was writing part of the next edition in his hotel. He announced to the hotel staff that I be given "special guest" status. The hotel was a bit of heaven on Earth for this young man talking—and writing.

I was invited by Colonel Marcos for a drive to Manila where he was arranging for a Lion's Club meeting at his hotel. He wanted to impress me with some big-shot cronies in Manila. Less impressive was his treatment of a boy at a motel on an overnight stop on the way to Manila. The boy was ignored when he asked to wash the Colonel's car. The next morning after a good breakfast we saw the boy again with the car nicely washed. The boy held out his hand, hoping for a tip. Marcos rejected the request and scooted him away. The boy got nothing. But I got something. I got all that I wished not to know about Colonel Marcos. I judge a person's worth not by their treatment of equals, but by their treatment of those lower on the social plane. None of the big wigs I was introduced to in Manila impressed me, and I was sad to see my host's shoddy treatment of the boy. This sentiment went deeper with Mahatma Gandhi, who said that the greatness of a nation and its moral progress can be judged by the way its animals are treated.

BJ and his family

BJ Bar in Pattaya, Thailand

As I say repeatedly, a great bonus of travel is the people you meet. My friend Huey advised traveler friends to include a necktie with their belongings. A traveler wearing a necktie experiences doors of opportunity otherwise missed.

Bangkok is an exciting city, cited often by travelers as their favorite international city. Nearby, on the warm welcoming Gulf of Siam is the city of Pattaya, where I met Bill Jones known as BJ, an American GI who fell in love with Thailand, married a Thai singer, and took residence there. He founded and operated the BJ Bar, and later the BJ Holiday Lodge. He and his lovely wife remain married after more than fifty years. A street in Pattaya is *Soi BJ*, is named after him. Snorkeling off the shore of Pattaya was awesome. I remember during an early dive saying to myself that I couldn't imagine more beautiful coral. Pattaya is a paradise to many visitors worldwide, with BJ a big part of its allure. BJ has met some of my friends who tagged along with me to Thailand over the years. These include David Vasquez, Helen Yan, Craig Dawson, nephew John Suchocki, and most recently, my wife Lil. BJ always has a smile for all. When asked about his popularity, he'll say it's simply being nice to people. I couldn't agree more.

A present-day very good friend is Dennis McNelis, whom I met in 1979 in Pattaya. At the time, Dennis was an American merchant marine from Somerville,

Bob Miner and Dennis

Signpainting in Pattaya with Dennis McNelis

Massachusetts. His seafaring duties were servicing US Navy ships in the south Pacific, which often meant months-long breaks between errands. His American friend who did the same was Bob Miner, who at the time was stationed in the Philippines. At the urging of Dennis, I connected with Bob on a visit there the same year. In December Bob and I were invited to a Christmas Eve dinner. Bob asked if I could eat dog, a specialty on the menu. Considering myself somewhat worldly and not hampered by a provincial upbringing I replied yes, it would be my first experience. Bob replied good, because the locals chose a dog who was an annoying pest that barked too much.

When seated at the dinner table, I saw sauce being poured on the raised leg of the dog while being openly roasted. When served, Bob began eating. No problem. But when I raised a bit of meat to my mouth my stomach retched. My stomach was telling me no, not to eat dog. I managed to get a bite in my mouth, but retching continued. I asked myself, who was in charge here? Me, or my stomach? I had no control of the retching. My stomach was in charge—the very stomach of my boyhood in provincial Massachusetts. (Years later on a trip to China I had my second dog-eating experience, but this time I was spared the knowledge of it.) After Dennis and Bob left the life at sea, they went on to successful careers back in the States—Dennis with

Koh Tao, Thailand

FedEx and Bob developed an online course in financial markets trading. They remain very close friends today.

While in Bangkok in 1983 I encountered Howie Brand, my dear friend from college days. World travel is Howie's hobby, and it was good to encounter him in exotic Thailand. After graduating from LTI, Howie worked in the defense industry for about nine years, developed and operated a small furniture business for fourteen years, and then became a physics teacher. Howie was one of the first to teach from the high-school edition of my textbook.

His last teaching position was at Suitland High School in Maryland, very near Washington, D.C. On a visit to his school in 1999 I was startled when he showed me a sign on the physics storeroom wall. An arrow pointed to *Conceptual Physics* and an opposite arrow to *Real Physics*. This misconception persists in many places. Howie became quite popular with his students and his enrollments swelled. Almost half of Suitland's students took at least one physics course. With expansion of the curriculum, officials decided that college-bound students could only elect "real physics," if any physics at all. In 2004 a resulting decline prompted Howie to pack his belongings, rent his home, and travel to the island Koh Tao off the eastern coast of Thailand.

He has lived there since. He likes getting around, so he travels to the United States or visits other countries about once a year. I see my dear friend Howie as a role model. It takes more than a bit of moxie for a retired teacher to get off their butt and "get away from it all."

A great friend I met in Hawaii was Walter Steiger who sat in my classes. We were elated to develop a mutual hobby—ceramics. We enjoyed our times together creating ceramic bowls at an art studio of the Community College campus of UH Hilo. Our bowls were nothing to write home about, but creating them was relaxing. The ceramic art instructor was another friend, Meidor Hu, who previously visited some of my classes in Manoa

when they were videotaped. Meidor comes from an amazing family of Chinese who immigrated to Hawaii years earlier when driven from their homes by Mao Tse Tung's armies. In 1990 when the Hu family planned and made a return visit to China, I was honored and privileged to be invited. The youngest of the Hu family, Tin Hoy, went on to be an information technology specialist

To China with Wai Inn, Ping and Meidor Hu

at UH Manoa, Mei Tuck became a medical doctor at PeaceHealth St. John Medical Center in Longview, Washington, and Meidor remains an art instructor at Hawaii Community College in Hilo. A most impressive family!

I visited China twice, first with Meidor's family, and twenty years later with Lil's family. The highlight during both visits was Guilin, with its famed Li River meandering through the awesome karst-mountain landscapes often featured in Chinese art. Views were best from the tourist river boats of the Li River cruise, that I took downstream in my first visit, and upstream in the second. That scenery is my favorite in the world and why it graces the cover of the fifth edition of *Conceptual Physical Science*.

In my world travels, the country I most often visited was magnificent Costa Rica, where my brother Steve settled in 1974. His home is located in a remote beautiful valley, about a forty-minute drive from San Isidro.

The Costa Rican Hewitts

In the early years it was a delightfully rustic home with hard-packed dirt floors, running water from a nearby stream, initially without electricity, with spaces near the ceiling for bats flying to clear up mosquitoes in the evenings. He married Elena, a local gal, in 1978, and soon little tots Stephanie, Travis, and Gretchen arrived to became a close knit family. We visited Steve for many summers. A festive Hewitt-Suchocki family reunion occurred in Christmas 1990. Fifteen years later followed a second

Cien Fuegos in Costa Rica

Happily married

family reunion. Thirty guests inhabited his newly constructed home near the top of the gorgeous valley. He calls his stunning home, Cien Fuegos. I like to call it the Taj Mahal of Costa Rica. And quite wonderfully, that was the site of our small lovely ceremony on Valentine's Day in 2005.

Contributing to the joy of travel were Ken and Joanne Ford cruising along the New England coast and entering the Gulf of St. Lawrence waters to Quebec. A more impressive cruise with them was aboard the Queen Mary 2, from New York to England. Six days later, upon reaching England, we departed in separate directions. After Lil and I traveled through much of Europe we visited Mona El Tawil-Nassar in her part of the world, Cairo, Egypt. She treated us to the same great pyramid tour that President Obama explored nine days later. Mona has "adopted" Lil as a "sister," and remains a close friend. Hooray to travel and long lasting friendships.

Dining with the Fords

In my travels to Kenya I visited a community of wood carvers and watched craftsmen skillfully carve a pair of large elephants, which I purchased and brought home with me to Broadmoor Village. Sorrowfully, the elephant pair was soon burglarized (which became the main reason to relocate to downtown San Francisco), I missed my elephants. A few years later a gift from Lil eased this loss. She knew that I was fond of elephant art, and crafted this forty-by-forty-inch hooked tapestry. It has graced the walls of our homes from then on. Thank you Lil!

SEVENTEEN

Retirement Years

J ust before the turn of the twenty-first century in 1998 I retired from class-room teaching. There's a strange twist to this. Before publication of my first textbook, students were the center of my attention. All my attention was on teaching. Students *were* my way. Weirdly, after my book was published with its need of continual revisions, students were *in* my way. My attention was redirected to the larger audience, those who learned from my book. I found it difficult to teach and write at the same time. As much as I loved the classroom, I retired at a still youthful age of sixty-seven to devote my time to updating the series of conceptual books—edition after edition.

My teaching colleagues arranged a retirement party at CCSF. My daughter Leslie planned the party. My office mate of many years, Jerry Hosken, was the host. Ron Ng from the hotel and restaurant department (who later inspired my grandson Manuel to become a top-notch chef) provided a sumptuous feast. It was a happy event sprinkled with stories by colleagues to honor me. One was Dave Wall, who entertained us with magic tricks

Office mate and host
Jerry Hosken

Dave Wall performing magic tricks

and spoke of the noisiness of my classes. He related how frequent bursts of student enjoyment could be heard outside in the hallways. The sounds of delight from there weren't a reaction to any jokes of mine, but simply a class response to understanding physics. Getting the best from our brains can be joyful. I was being honored for being a teacher to elicit that joy. That honor was extended to renaming my favorite classroom, S100, the Paul G. Hewitt Physics Lecture Hall.

Many family members and friends attended this wonderful party. Sister Marjorie mingled nicely with all, as did Huey and Sue Johnson. Millie, my wife of more than forty years yet separated earlier, chatted with many mutual friends and some new ones. She met a new friend, Lillian Lee. Millie told me privately she liked Lil a lot, and to be good to her. In 2004 Millie's heart failed and she passed away. A year later, I married Lil.

Me and my love in Japan

Retirement offers time for reflection. Looking back, I think of the different homes I occupied over my teaching career. The first was in Broadmoor Village neighborhood of Daly City and very near CCSF. In the 1970s, when my family spent a lot of time in Colorado, many guests at our Broadmoor home included Roger Gribble with his iguana, artisit Mary Jew, photographer Jim Morgan with Judy Johnston, Sharon Lewis, and Tim and Rose Gardner. Joyful parties were frequent. My happiness there was interrupted by a string of burglaries that transformed my much-loved home to merely a house. My colleague Jim Conley convinced me to relocate to an apartment in downtown San Francisco. After more than a month of persistent searching door to door, I was lucky to find a great two-bedroom apartment with two indoor-parking spaces in North Beach neighborhood, situated between Coit Tower and the Transamerica Pyramid—One San Antonio Place, 2D. I rented that apartment for twenty wonderful years.

Shortly after moving into my North Beach apartment, cousin Bud Baruffaldi came up with an idea for his son Robert, who was unhappy working in the family Concrete Restoration Company in Massachusetts. Bud said to Robert, "If Uncle Paul can do physics, so can you." Agreeing with Bud's assertion, I invited Robert to come to San Francisco and have a go at it. He occupied the second bedroom of my apartment with its own bathroom. Robert was a perfect housemate. Never in those four years did I wish he weren't there. Robert became a best friend to son James and also a dear friend to daughter Leslie, who at the time lived in an upstairs apartment. Robert and Leslie both attended CCSF for two years and then transferred to San Francisco State University for another two years. In accord with his dad's hopes, Robert graduated with a physics degree. Robert continued to UCLA and earned an MS in atmospheric science, followed by

James and his cousin Robert Baruffaldi

a government career as a meteorologist at the National Weather Service in Sacramento. Leslie's degree was in geology. As discussed in Chapter 13, Leslie joined me and her other cousin John Suchocki in creating our first physical science textbook.

David Ferris

A bonus of the North Beach rental was meeting upstairs British neighbor David Ferris. David spent part of the year living in London where he grew up, and part in San Francisco. He was a pioneer in the computer revolution and made more than a good living helping large organizations and computer firms in California plan their technology strategies. He relates an interesting story. After he dropped out of Stanford University, he and a pal both overstayed their welcome at a neighborhood coffee shop. The manager asked them to leave, never realizing that he booted from his premises someone destined to become the wealthiest man in the world—Bill Gates. Lil and I rate David the most interesting of our friends. His present chum and former wife Jean from London are the most regal of our friends. And recently adding to this duo is their son Nicholas and wife Vinaya, and their toddler daughter Sienna—all, very impressive neighborhood friends.

As part of life after retirement, Lil and I joined our lifelong friends Paul and Norma Ryan in Florida part time. We settled in the charming city of St. Petersburg in a condo on Tampa Bay, twenty minutes away from the prize-winning Tampa International Airport—a must for travelers. Our first new friends were neighbors dentist Jerry and nurse Barbara Donahue. Physics author Ray Serway and his wife Betty became friends and neighbors for a couple of years before they moved back to their home state of Virginia. Likewise with English physician Mike and nurse Jane Jukes, the spunkiest of our Florida friends who lived both in Florida and Maine. Lil and I similarly enjoyed our routine of living more than half a year in Florida and the rest of the time in San Francisco with Lil's mom.

More sunballs

Mom and sunballs on a sidewalk

During my retirement years I was delighted to present guest lectures at Florida high schools. A favorite is the Burl Grey story of learning physics on the painting scaffolds in Miami. Other favored topics were satellite motion, how sailboats are able to sail upwind, nicely showing the usefulness of vectors, and most amazing to students, *sunballs*, the circular images on the ground beneath sunlit trees cast through "pinhole" openings between leaves above. Since these solar images escape the notice of most people, it's nice to be the teacher to introduce what people have always looked at but never seen. And who doesn't love rainbows? Due to rainbow geometry, each rainbow has its center aligned with the viewer's eye, making that rainbow personal for that viewer. I've always felt strongly that an overview of physics should

encompass a wide swath that is well-paced to progress from mechanics to electricity, magnetism, the delights of light . . . and culminate with rainbows!

Our new Florida location made it easy to drop in on Jacque Fresco and his partner, Roxanne Meadows, who live in the small town of Venus in lower central Florida. They created a several-acre model community for future living, with dome-shaped quarters and small laboratories, and even a heliport for helicopter visitors. Jacque had earlier steered me into a life of reason and science. On one visit Lil and I brought our friends Paul and Norma Ryan. And what did I see? Unsurprisingly, Fresco, being a "born teacher," sharpening the thinking of my

With my mentor, Jacque Fresco

guests by teaching them how to draw objects with proper perspective. In all my life I have never met a more inspiring person so dedicated to teaching.

One of Fresco's many projects was a book he was writing. He asked me for suggestions and edits. Oops, I found a glaring one! In discussing aircraft, he boldly stated that he never believed Bernoulli's principle. Whoa, whereas

A lesson from Fresco with Ryans and me in Venus

Bernoulli's principle is commonly misunderstood, even by teachers, it's a physics cornerstone in fluid dynamics. For Fresco to say he didn't believe it would be a red flag to those knowledgeable about the physics of fluids. I explained to Fresco how and why Bernoulli is misunderstood, and strongly suggested he instead correct his earlier *misunderstood* Bernoulli's principle. Was he okay with my suggested change? Sadly, no. He would have none of it. I was heartbroken to find my hero "set in his ways," he had no interest in correcting his misunderstanding. This bothered me.

In my profession of explaining physics, I profited greatly by the many times I was confronted with being wrong. I then gleefully corrected my errors. Being able to admit to mistakes and correct them should not be rare. My knowledge of physics has never stopped growing, in part because I continue to learn from my mistakes. It's nice to feel secure enough that a valid criticism is welcomed rather than being rejected out of hand. We should be comfortable telling our students, "I don't know, let's see what we'd need to find out."

Burl Grey

I never tired of telling the Burl Grey story about learning physics when we painted billboards together. I relished communicating my excitement in learning how physics principles account for so much in the everyday world. However, I never shared with my students something very important to my life—that my shift from religion to science began on those scaffolds. Those conversations with Burl led me to give up on my belief in a personal God. I kept private the emotional challenge that came with accepting the loss of an all-seeing protector, no notion of justice prevailing in an afterworld, and that life's purpose wasn't in the hands of a supreme being, but in our own hands. But everything then seemed simpler with the absence of superstition, original sin, and damnation. The unknown became much less fearful.

I classified myself as an agnostic rather than an atheist, to keep a personal door open to some unknown realm outside the domains of religion and science—to some sort of universal consciousness less provincial than espoused by the world's great religions. I certainly don't value the idea of indoctrinating children before they can reason. Nor do I wish tending toward a jealous deity

I'd rather hang out with friends who have reasonable doubts, than ones who are absolutely sure of everything.

obsessed with being worshipped. Yet my Darwinian thinking remains stymied in imagining such things as the evolution of RNA becoming DNA and so on without some *guidance*—something very vast—or, delightfully very simple. Who knows? The mysteries that accompany our journey through this enormous universe await discovery.

Years earlier, I encountered a handicapped man at a Salvation Army store who was looking forward to an evening's religious service. I wondered if I should announce to him that freedom could be attained by rejecting beliefs in a god. But could he withstand the emotional shedding that I knew to be initially quite painful? And what good would that do him? What of the comfort of feeling connected to something vast and meaningful? As my sister Marge says, religion's better focus is human community. She notes that the core principles always boil down to compassion.

Albert A. Bartlett

As I revise my textbooks, I'm grateful for friends whose ideas and insights have helped me write better editions. One such person is Albert A. Bartlett, of the University of Colorado at Boulder, who boldly asserted, "The greatest shortcoming of the human race is our inability to understand the exponential function." Unchecked growth can be disastrous. Many of my books have an appendix, *Exponential Growth and Doubling Time*, which reminds us of the dangers of unchecked population growth.

The world population when I was born in late 1931 was about 2 billion people. In 1974 it doubled to about 4 billion. That's a doubling time of about forty-five years. And some forty-five years later in 2023 it doubled to 8 billion. So in my lifetime the world population has doubled twice! Today's vastly greater number of people since my childhood is very noticeable! A narrative that still prevails is that growth is good—even population growth. On a personal level, I think of the unchecked growth of obesity, or worse, cancer. Count me out in any celebration of reaching the 8 billion mark for population. And cheers to Al Bartlett.

Evan Jones   David Kagan   John Hubisz and Chuck Stone   Jeff Wetherhold

Another professor friend is Evan Jones of Sierra College, who helped me clarify my treatments of Bernoulli's principle and LEDs. Other resourceful friends include David Kagan, John Hubisz and Chuck Stone, who each made significant contributions to my writings. Lastly is Jeff Wetherhold, who kindly gave me his father's blueprints that verify an idea that I wanted to write about: railroad trains are kept on tracks via the slightly tapered rims of its wheels.

My retirement years have given me time to reflect on the many former students who became lifelong friends. The list begins with the five sons of Louis and Phyllis Vasquez, colleagues who taught physical education at CCSF. My first student was Michael in the fall semester of 1967, then David the following semester, and Rodney the next, then John, and finally Bob. These five brothers went on to careers in education. If David weren't in the mix, the videos that feature my classroom teaching may never have happened. Recall from Chapter 11 that David first brought video cameras into my 1982 CCSF classroom, which led later to the Hawaiian set of thirty-four videos,

Vasquez family of teachers

Conceptual Physics Alive!. Other students include Calvin Pon, Marissa Milan, Sandy Sandoval, Ron Ng, Katherine Lee, Freda and Mac Richardson, Paul McNamara, Hideko Limogan, Susan Jacobson and her son Toby, Cuong Ho, Alexei Cogan, Tenny Lim, Suzie Tan, Sidney Keith, Helen Yan, and, more recently, Gordon Ma. Topping this list of friends at CCSF is Lillian Lee, now my wife! Interestingly, Lillian never took my conceptual course, but instead opted for the algebra-based physics course required for her pharmacy career.

I was most fortunate that CCSF supplied me with adequate numbers of teaching assistants (TAs). Having multiple TAs nicely enabled my policy of re-take exams for Physics 10 students. The earliest TA that I remember was Lillian Tham, followed by Roger Gribble and Denise Davis. When I wrote my first book in 1969 my TA was Lynette Fung. Soon after was Mary Jew, then Katherine Lee. Over the years TAs progressed to Marissa Milan, Sandy Sandoval, Ken Sherey, Mac Richardson, Lillian Lee, Tenny Lim, and Ann Chang. Later came Susan Gross, C. J. Rember, Alexie Cogan, and Glenda Ginn. I list those whose names come to mind. There were quite a few others. Quite wonderfully, I was honored to be master of ceremonies to two TA

With JoJo Dijamco and Stella Chan

Talented daughter Abby Dijamco

weddings—first Katherine and Victor Ng, and some years later Tenny and Mark Clark.

The last of my TAs just before retirement was Stella Chan, a deeply religious student in her late twenties. When she told me that she was without a boyfriend, I suggested she find one at a church function. She resisted, not wanting to use her church to advance personal wishes, a no-no. I advised her to consult her minister about that. The result was predictable, for in short order she found Jojo Dijamco at a Friday evening church social. Wonderful. A year and half later, when I was in Hawaii, I flew to attend their wedding in San Francisco. Stella later gave birth to three lovely daughters, nicely featured in my books. It's nice when good advice is taken.

The most impressive friend I gained in my CCSF years is Will Maynez, who wears three hats. One is for being the CCSF physics lab manager for 33 years, and the second is being the curator and storyteller of the Diego Rivera mural. The third is being an accomplished playwright. Will is truly a man for all seasons. The Diego mural known as Pan American Unity is a world-class treasure painted by Diego for the 1940 Golden Gate International Exposition on San Francisco's Treasure Island. Its home after the exposition was at CCSF. In 2015, Will and CCSF librarian Julia Bergman were recognized by The Art Deco Society of California for their work in being stewards of the mural. Realizing the mural could last centuries longer than the CCSF

Julia Bergman and Will Maynez

building it inhabited, Will initiated studies about feasibility of moving it. Soon the SF Museum of Modern Art (SFMOMA), "borrowed" it for several years. Its temporary site in downtown San Francisco has given it a worthy international exposure. More on this can be learned via Will's newsletter on the mural's website: riveramural.org.

> There is a pool of good. No matter where you put in your drop, the whole pool rises.—*Will Maynez*

Most of these friends appeared in the photos in the many editions of my books. The front matter of various editions list them in the Conceptual Physics Photo Album. If it is said that my books are photo albums of people I admire, I'll agree. True enough. Happily, my publisher has never discouraged that. I credit some of the most competent people I've met over the years— those who toiled to produce the many editions of my books. Publishing *Conceptual Physics* started in 1972 with Little, Brown and Company, which was then acquired by Harper & Row, followed by HarperCollins, then in succession, Addison-Wesley, Prentice Hall, , and finally Pearson Education. The publishing industry has seemed to me to be a game of "musical chairs." It worked well for me. How lucky I have been supported by teams of talented publishing people.

Of particular note in later editions are photos of children supplied by Angela Hendricks, the granddaughter of my brother-in-law George Luna. Angela is an amateur photographer and teacher in Pocatello, Idaho. I'm delighted to show some of my extended family photos that have nicely

Angela and
Jake Hendricks

Like everyone, I'm made of atoms—which are so tiny and numerous that I inhale gazillions of them with each breath. Some stay for awhile and become part of me. When I soon exhale them, they're breathed in by other people and a tiny bit of me becomes part of them. Air in the atmosphere quickly recycles, which means that atoms I inhale include atoms exhaled by people from all over the world, who then become part of me. Wow, in this sense, we're all one!

Tiny Georgia Hernandez

Itty bitty Kara

A nice snug fit in this water warmed by the Sun! Water is also warmed by radioactivity in Earth's innards that heats thermal springs and geysers. How nice that nature supplies warmth from the Sun above, and by radioactivity below. Yum physics, both ways!

Although the temperature of these sparks exceeds 2000°C, the heat they impart in striking my skin is very small -- which illustrates that *temperature* and *heat* are different concepts. Learning to distinguish between closely related concepts is the challenge and essence of *Conceptual Physics*.

Little Terrance Jones

Isaac Jones

graced in her Uncle's books. I credit the diversity portrayed in my books to Angela's photos. My favorite is little Georgia Hernandez citing the role of atoms in her body. Angela's aunt Corine's first-born son, Terrence, holds a fireworks sparkler in an early edition. Years later, Terrence's son Isaac held sparkler! A West-coast family member is my granddaughter Kara Mae Hurrell talking about some yum physics while in a bucket of water! Cheers to family!

When I married Lil, I was welcomed into her huge family of cousins, and friends. Lil was born in China in 1958 during bleak times due to policies of Mao Tse Tung. In search of a better life her family left their home in

Lil with her mom and dad   Later with amazing Mom

Lillian's family

Canton to Hong Kong. Shortly thereafter they continued on to Vancouver, B.C., Canada. With an uncle's sponsorship, they later immigrated to San Francisco. Lil's dad was a self-taught butcher in China and Vancouver, and he became a cook at prestigious Kan's Restaurant in Chinatown. Lil's mom worked as a seamstress.

We're dealt a deck of cards at birth. The quality of our lives relates to how we play that deck. Much is enhanced by the skills we learn. In all cases, however, we must acknowledge that we have zero control over the biggest factor in playing that deck—*luck*. I've been blessed with the best of good luck, with a large blessing being married to the gal who joined me in my writing efforts. Happiness in life is more than what we've done or where we've been— it's mainly *who you are with*. Being with all the yum people I got to know at places of teaching and professional meetings contributed significantly to this old man's amazing life.

Hooray to satisfaction and joy. I see happiness composed of having three ingredients: having someone or something to love, something meaningful to do, and something to look forward to. I am lucky to have all three. When Lil was my teaching assistant in 1978, I had no idea she would one day be my wife. My first wife, Millie, and then Lil, have given me all three. Another adage tells us that people most value recognition and appreciation. I've been extremely lucky to experience these as well. For me, there can be no better life than teaching—being the one to guide student's minds to the connections of nature—in particular the laws of physics. As teachers, we lead our students to a scientific appreciation of nature's magnificence. What could be more satisfying?

I wish to share a personal tip that injects happiness into a relationship, which I think is worth sharing with your friends. Lil and I maintain harmony by a "get-it-off-your-back day" once a month. On the first day of each month we make an effort to correct problems before they grow. As an example, when Lil called attention to my habit of leaving coffee grounds on the counter instead of cleaning them up, I corrected my behavior in the spirit of "What can I do to make her happier?" Occurring only one day a month avoids bickering, that if overdone frequently becomes nagging. As the saying goes, "It's not fires or hurricanes that bring mighty trees down—it's the tiny beetles." Having a routine that addresses small issues prevents larger ones. And very importantly, we're simply nice to each other. Isn't the key to a good life simply being nice to others?

Since retirement Lil and I devoted ourselves the thirteenth edition of *Conceptual Physics*, all the while creating new Figuring Physics columns that appear monthly in *The Physics Teacher*. We do the same for the nine-month column, Focus on Physics, *The Science Teacher*, a journal published by NSTA. I do the writing; Lil edits what I do, and colors all the artwork. Work and play keep us delightfully busy.

A very nice offshoot of retirement has been attending the monthly CCSF Lunch Bunch gatherings of retirees when we're in San Francisco. A dozen or

CCSF Lunch Bunch

so retirees meet at local restaurants for chatter and camaraderie. This memoir includes feedback from some of them. One is Karen Grant, a student in my class when my fingers were broken in the anvil-sledgehammer demonstration of inertia. That was something to talk about! English teacher Eloise Rivera provided good suggestions. Another feedback contributor is interim CCSF president Jackie Green, who seldom missed a meeting. Then there was feedback from the physics guys: Dan St. John, Norm Whitlatch, and Will Maynez. The Lunch Bunch host, Peggy Vota, advised me to be sure the conclusion of my memoir tells how I'd like to be remembered. I'd like to be remembered as a teacher who helped students discover that physics is a part of their everyday lives, and who helped to change the way physics is taught nationally, and to help create a national physics-first movement. And most of all, as a teacher who placed love along every step of the way.

Do I have advice in this memoir to the reader? Yes, of course, I'm a physics teacher! My advice to whom? Is the reader a physics teacher? Maybe a beginning one, or presently in the trenches—or an elder like me? Or someone outside of physics or teaching, just taking a peek at what this old man has to say. Knowing that one size seldom fits all, I'll keep my advice general. To begin with, whatever your stage in life, follow your *passion*. Love and excel at what you DO, for how well its done will likely define you. Become your own expert in better understanding what's in your world, whether its learning about ants, plants, or those invisible fearful bio critters on our horizon. You may find that learning can be exhilarating, assuming that the curiosity you had as a child is still part of you. If it's sleeping, awaken it! Be like the many who are glad about what they do. Success will be more than your possessions, or where you go. It will involve *who you're with*. Hang around people who have something interesting to offer. That I hung out with people like Frank Oppenheimer, Albert Baez, and Ken Ford very much contributed to my accomplishments. Above all, be kind to others and yourself. Just be nice. We're all a part of this human adventure. Relish that!

EIGHTEEN

Attitudes, Opinions, and Hopes

I am sharing some of the attitudes, opinions, and hopes that have influenced much of my writing and teaching. Please bear with me when we're not "on the same page." Unless you've walked a path parallel to mine, you'll not fully share my views. I hope we can agree to disagree at times.

I think personal attitudes are central. The Army slogan, "Be All You Can Be," goes further than soldiering. Recall in Chapter Two the wisdom of defeated boxer Tony Karas who said: "There's always a better man—and I met that better man tonight." People with Tony's attitude usually go on to have happier lives than those who succumb to grumbling. Cheers to a healthy attitude.

Becoming a physics teacher at City College of San Francisco was an enormous achievement. Having a career to share my love of science with others—many others—became my personal heaven. Yet a friend and fellow physics teacher had a very different attitude. He considered himself a failure because he was *only* a physics teacher at a community college. Guess who was happier in their teaching. I can't say it enough: *An upbeat attitude is everything.*

The experiences we encounter are important. As important, if not more so, is our *response* to them. As said in Chapter Four, being drafted into the Army was a mistake. Due to being significantly underweight I shouldn't have been drafted in the first place. But luckily not having to experience combat, I benefited greatly from the discipline, camaraderie, and improved physical fitness it provided. Rather than being resentful and having a miserable time, I made the best of it. Similarly, in becoming a community college physics teacher being thrilled with that success. Perhaps I view the world through rose-colored glasses. It seems to account for my positive attitude.

And then there's my wife Lil. When I first got to know her, she let me know that she felt pretty okay with herself. As she put it, she wasn't the prettiest or the ugliest, but was in between. She wasn't the smartest or the dumbest, but was in between. She never let herself become consumed with trying to appear to be better than she was. Some lipstick at times, and a minimum amount of makeup. Okay, she colored her hair at one time. But she never even had her ears pierced. Of course she likes looking good—but she was never *obsessed* with it. I fell in love with my "in-between" Lil. That love per-sists today.

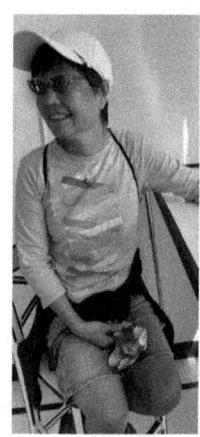

Lil

It's said that early impressions are often lasting ones. I remember my first tour of the CCSF campus when I read the inscription at the entrance to the Science Building: "The Truth Shall Make You Free." This was quite welcoming, for the pursuit of truth guided me into physics. Unlike other fields of study, physics is the same worldwide—physics in one part of the world is the same physics as everywhere else. Stability is nice. However, truths vary over time—even in physics.

Truth in general? Isn't it the label we affix to our personal beliefs—most instilled in us by our parents. They define our identity, which is fine.

CCSF Science Building

However, I'm somewhat bothered by many people with newly acquired beliefs having the strongest opinions about matters they least understand. Some are emotional and often conform to a defining narrative. As a teen in church, I remember standing with the congregation and reading aloud the Nicene Creed—a declaration of faith with a list of statements deemed to be true. I thought many were questionable, but to belong to the Congregational Church in Saugus, I had to profess belief in all of them.

Isn't this the case with *tribes*—whether social, political, religious, or whatever? To belong to a tribe you must accept its narrative. Each tribe has its own version of truth. Some "truths" can be outgrown with time, and some not. "Truths" ingrained in children can become moral straitjackets never to be outgrown. I have always felt lucky that my childhood was without intense indoctrination. That enabled me to be open to and influenced by Burl Grey and Jacque Fresco in Miami. Years later Burl told me that I was the only fellow painter to pay heed to his eccentric opinions, which likely threatened the identity of other painters. Hooray to being free of childhood indoctrination!

The appeal of tribes is understandable. People want identity—to "belong" with others in the quest for meaning and purpose in life. For centuries, churches and armies rose to that occasion. I benefited greatly from belonging to the Army with no combat and also from the schools I attended.

But I'm bothered by the downsides of tribes: mainly their obsession with growth and the quest for power that comes with swelling membership—of pushing new-found freedom into a power movement. I'm fond of individual people because most are decent. I can't say the same for groups who churn social divisiveness. Fortunately, not all tribes harbor this downside. The one group that I'm satisfied belonging to is the human race—warts and all.

When thinking about tribes there's the story of the small group of people in a desert confronted by threatening tribesmen on camels, who with guns raised asked, "Are you Christians or Muslims?" When the answer was, "We're tourists!" the tribal leader said, "Okay, you may pass on." In times where tribes are becoming more common and violent, I like being a tourist. I benefit from the tourist stance by not judging people by their religion or politics, which most often merely reflects their information sources. Whatever way they lean, most people are decent.

Most of my life has been enhanced by decent people—from those I hung around with in the boxing crowd, the Army, and the thousands in student classrooms. I have pen pals. The longest running is Suzie Tan, my student in 1981. Another more recent pen pal has been Gordon Ma, a student in 1995.

An especially interesting international pen pal is Einstein Dhayal from India, who began corresponding with me at age thirteen. He shared unrealistic speculations about physics to which I responded that he should first study and learn elementary basics well and not waste time with poor knowledge. Our correspondence continued for years. Today he is my protégé on his way to earning a PhD in physics in America. At age seventeen he was accepted with a scholarship at Michigan State University. To smooth acceptance to new college friends, Lil suggested a nickname, Ed, the initials of his name. How nice that the nickname wasn't needed, for he gets along wonderfully with others. He visits us

Chatting physics with Einstein Dhayal

yearly and has become our "adopted nephew." Lil and I both love our young physicist.

One of the habits that I find enriching is touching base weekly for family chatter with my siblings, Marge, Dave, and Steve. We refer to this as "Sibling Sunday," initiated a few years earlier by Steve. Steve, having been a mariner most of his life likes his nickname, Boatz. In childhood I benefited from my close family; as a teacher I strove to create a family atmosphere in my classrooms.

I now wonder about the world of tomorrow. If we heed the many dooms-day scenarios, it will be pretty bleak. The media prioritizes the negatives. We don't hear people reading and writing more than at any time in history, people working fewer hours per day with fewer injuries, and traveling more, all with greater political freedom. People today are much healthier than in the past and live longer. We innovate and adapt to changing conditions. Humans are good at adapting.

Being with Aiden Hewitt

I think of the future that awaits my young great-grandson Aiden, son of Manuel Hewitt. He'll likely never know cancer, diabetes, mental illness, and other diseases we presently contend with. I think it's a safe bet to say he'll most likely live beyond a healthy one hundred years old. One reason for my optimism is the recent change in the direction of youth. When I graduated from college, the world had just been through a world war with the scare of another on the horizon. Too many of my classmates found employment in the lucrative arms industry. Since then, I'm in awe of the greater numbers of students majoring in computer science directed to the medical field—a desirable shift from conquering people to conquering diseases. Cheers to healthcare over warfare. Humans are still evolving.

Many of our opinions have been shaped by those of others. While at Lowell Tech I was quite influenced by the nineteenth-century physicist Lord

Kelvin, who famously stated, "When you cannot express what you're speaking about in numbers, your knowledge is of a meager and unsatisfactory kind—you've scarcely in your thoughts advanced to the stage of science." I found Kelvin's statement profound when I began my quest to learn more about science. It evoked spirited discussions with my classmates. His statement has guided my thinking. Numbers *are* important.

I also stressed in teaching the importance of the word *some*. Consider the statements "Molecules travel faster in hot water than in cool water." "Forces applied to objects change their speed." These statements are questionable. It's more accurate to say that *some* molecules, *some* forces, and so forth. The importance of qualifying statements extends to the social realm. "Older people are wiser than younger ones." "Students at UC Berkeley are wealthier than students at CCSF." Better to say *some* older people, *some* UC Berkeley students. Let's be aware of the word *some* in general statements. It's important. Some of us love learning physics. Some of us don't. In the real world there's always some of this and some of that. Wisdom is considering both.

I think of wisdom when being told of an old man standing alone on a busy street corner holding a sign in protest. When onlookers told him that he couldn't change the world with that sign, the old man replied, "I'm not trying to change the world, I'm trying to keep the world from *changing me*—this is what I believe in."

What we believe, however, is often influenced by others. In politics, for example, policies that are a mix of good and not-so-good are presented to the public. Rather than dealing with trade-offs, opponents ramp up news coverage that exaggerates minor downsides. Policies are then trashed and deadlock prevails. This has always bothered me.

And then there are hopes. A great hope for an expansion of physics taught in high schools was boosted with my meeting a personal hero, Nobel laureate Richard Feynman. In the Introduction to this memoir recall that I spoke of meeting him at a 1987 AAPT conference in Pasadena, where the topic was

Richard Feynman

"What kind of physics should be taught in high school?" As a panel member, I took issue with the high status accorded to physics in high schools, presumably due to the rigor entailed. I posed a different order; the most rigorous course could be basket weaving. That's right, basket weaving—where its final exam is creating a good-looking leak-proof basket into which water is poured. If the basket leaks, the student fails the course. This, instead of physics, would be the school's "killer course." My point was that degree of rigor in a course need not be due to the subject matter, but to the depth of the academic plow controlled by the teacher. A first physics course ought to be a delightful learning experience that would bring physics into the educational mainstream along with writing and arithmetic. Rigor would be welcomed by students inspired to take a second physics course. Feynman was okay with all this, but then asked where the new qualified teachers would come from to serve "physics for all." My reply was that if we made the introductory course more captivating, more students would be drawn to physics, then major in it, and in ten years join the ranks of needed teachers.

Ten years later at a physics meeting in Las Vegas, a young man stopped me in the hall to tell me how his encounter with my book had steered him to major in physics and then become a physics teacher. I thought to myself, "Did you hear that, Richard?" This was but a sample to verify that a conceptual orientation of physics in the present time could help increase the ranks of new physics teachers in following years. I'm happy to say that the physics-chemistry-biology sequence has nicely caught on in high schools. My hoped-for "physics first" became commonplace.

Nevertheless, there remains a shortage of high-ability physics teachers in America. Some altogether new teaching method beckons. As effective as my lectures were, I often had the eerie feeling that I was among the last successful classroom lecturers. I felt I was part of a shrinking minority. Some better means of imparting information, and even inspiration, would come down the road and relegate the classroom lecture experience to history. And here we are, on the threshold of that new age: AI! I remember how pocket calculators that emerged in the 1970s were not welcomed in school classrooms.

Their downsides were exaggerated. Is this the case with AI? Shouldn't it be viewed through a lens of opportunity, rather than fear? AI may elevate the way learning is accomplished. Hooray! Yet there is reason for apprehension. By way of trade-offs, will upsides be greater than downsides? My hope is that the different classrooms of tomorrow be very much better. (For the record, this memoir was written solely by me, NOT with the aid of AI.)

My life as a physics teacher has indeed been awesome. I cannot imagine a more worthwhile and satisfying career. A big reason that teaching fit me so beautifully relates to what I've said previously: The quality of one's life depends on the kind of people one hangs out with. My time is best enjoyed with curious-minded people who share my love of physics. In my student days, that occurred in rap sessions with fellow students. After becoming a teacher I preferred to be with my colleagues both at CCSF and professional meetings—birds of a feather. Not to be overlooked is the largest pool of curious-minded people ready to learn physics—our students! Many reminded me of myself in earlier years, so no wonder I loved teaching young people that I identified with. How joyful it was year after year to guide fertile minds toward viewing their everyday worlds through my much-loved lens of science—focusing on physics. Much of their appreciation, and mine, was the assurance that what they learned was both relevant and valid.

Looking back over the decades of teaching and writing, I see that *Conceptual Physics* was unique. Although I was not motivated by money to create it, I certainly welcomed the royalties. The first check transitioned me from being always in debt. Continued royalties provided financial comfort for half a century, allowing me to cover college expenses for relatives, assist in the purchases of homes for family and others, and be generous in those in need. Having no appetite for a lifestyle of owning expensive items such as boats or airplanes, and not into alcohol or drugs, textbook royalties did more than meet personal needs. That wealth was appreciated.

On February 8, 2021, I suffered a stroke in our Florida condo. How fortunate that Lil was nearby and immediately called 911. Luckily, an ambulance was in the neighborhood and quickly whisked me to nearby Bayfront

Hospital where I was greeted and treated immediately by emergency staff. Their high-level competence was followed by excellent home care by physical and occupational therapists. Whew! Thankfully, within a year I recovered completely. Let's all hope this high-level medical care will expand to all.

My heartfelt hope in writing this memoir is that I've inspired some readers to become physics teachers—ones who in addition to teaching good physics will rev up a love for learning itself, placing drops of good toward a better world.

My experiences in teaching and traveling have been awesome. More awesome is simply being born. The odds of a human being born are so minuscule that being alive is the highest prize of the universe. But that prize comes with a certainty—eventual death. Before birth, our experience of the universe and our place in it was blank. With no consciousness there was nothing. If that blank state is our fate after death, then there's nothing to fear. If, however, there's more, hooray! But there needn't be. Death as the price of birth is a fair cosmic bargain. Along with grieving the loss of loved ones, we should celebrate that they knew life. As for my remains when I die, I'm planning on an environmentally friendly *green burial*—no embalming, no formaldehyde, a shroud instead of a casket, and a welcome feast for all those teeny-weeny munching creatures that will return me to the soil.

Christmas 2022

ACKNOWLEDGMENTS

This memoir is an extension of keeping a diary that began in my teens, and notes that I jotted down now and then after retirement. My notes had no purpose, other than perhaps being of future value for my grandchildren and my wife Lil. I shared some of them with Gordon Ma, a new friend who was a student in my Physics 10 class in 1990s. Gordon's response was ecstatic, citing some of the stories he found humorous and insightful. He strongly advised that I share my writings with a wider audience. Huey Johnson had earlier suggested I write a memoir, and together with Gordon's urging, I took the project more seriously. Lil agreed. She had been critiquing my textbooks for decades, and did the same with this memoir—nicely designing the interior with its many photos.

When I shared segments of chapters with others, many responded with valued suggestions. Foremost were Ken Ford, David Kagan, Chelcie Liu, Fred Myers, Bruce Novak, and Chuck Stone for all had previously contributed to my textbooks. For his encouragement and support I thank my nephew John Suchocki. I am also grateful to my family for supplying photographs. A tip of the hat to dear friend Judith Brand for her edits. For a final polish of skilled copyediting I thank Academic Consulting and Editorial Service (ACES) in India. Finally, my deep gratitude to World Scientific Publishing and their editors Yong Qi and Rhaimie Wahap for bringing the book to publication.

My Family

Dad and mom visiting Florida   Daily boxing on the porch in 1947

Mom with her grown-up kids

Four siblings
in 2020:
Dave
Margie
Steve
Me

A California hippie family

Daughter Jean Hurrell's family on Salida porch

Son Paul's family

Leslie with my granddaughters
Megan & Emily Abrams

My Lil's Family

Me and Lil

Smiling Dad

Delightful Sneezlee

With sister-in-law Linda Sinn
—the best of the best

Strolling with Mom

A Lee family celebration

Some Relatives

Niece Cathy
and Bill Candler

Sister Margie
with daughters
Joan Lucas and
Cathy Candler

Nephew John Suchocki's family

With brother Dave's
daughter Nancy Hewitt

With brother Dave and Barb

Manuel and
Aiden Hewitt

Hugging niece
Gretchen Hewitt

Brother Steve's family

Millie, Paul, James with Idaho George Luna family

George's family with spouses grown up

Both Luna Hewitt gathering in Salida

Some Friends

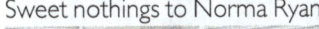

Lori
Patterson

Marissa Milan
and Sandy Barry

Tenny Lim

Sweet nothings to Norma Ryan

Toasting with Tsing Bardin

Mike and
Jane Jukes

Dennis McNelis' family

With Bob Miner's family

With Nathan and Calvin Pon

My 90th birthday celebration—Will Maynez, Dave Vasquez, and Marissa Milan

Helen Yan, Will Maynez, David Yee, Paul Hewitt, Jill Evans, and Rosa Alvis celebrating the 90th Birthday of City College of San Francisco.